A ONE-DIMENSIONAL
INTRODUCTION TO

CONTINUUM
MECHANICS

T0320620

A ONE-DIMENSIONAL
INTRODUCTION TO
CONTINUUM
MECHANICS

A. J. Roberts

Professor of Applied Mathematics
University of Southern Queensland

World Scientific
Singapore • New Jersey • London • Hong Kong

Published by

World Scientific Publishing Co. Pte. Ltd.

P O Box 128, Farrer Road, Singapore 9128

USA office: Suite 1B, 1060 Main Street, River Edge, NJ 07661

UK office: 73 Lynton Mead, Totteridge, London N20 8DH

Library of Congress Cataloging-in-Publication Data

Roberts, A. J. (Anthony John), 1957–
     A one-dimensional introduction to continuum mechanics / A. J.
Roberts.
          p.   cm.
     Includes bibliographical references and index.
     ISBN 981021913X
     1. Continuum mechanics.   I. Title.
QA808.2.R63   1994
531--dc20   94–30315
                                                                CIP

Printed in Singapore.

Preface

This book was born out of my desire to introduce the fascination of the dynamics of continuous media, such as air and water, to students at an early level. I am especially keen to explore with students the application of mathematics in the world that we see and feel around us.

Most introductory treatises on continuum mechanics "dive into the deep end" of the three-dimensional dynamics of air and water; while a worthy aim, this does have the difficulty of introducing important and basic modelling concepts at the same time as requiring students to use only recently learnt abstract tools such as multi-variable calculus. Typically, the three-dimensional continuum equations are then simplified to show the dynamics in a variety of simple situations—simple often because the dynamics are specialised to one dimension. Indeed, it is amazing how many important models of physical processes are based entirely within the dynamics of a one-dimensional continuum. This book approaches continuum mechanics entirely within such a one-dimensional framework. It can be understood without any multi-variable calculus, and needs only an elementary introduction to partial differential equations (such as the technique of separation of variables). Despite this simple base we discuss the dynamics of vital physical processes such as algal blooms, beam bending, blood flow, tidal dynamics, dispersion in a channel, and the greenhouse effect.

The plan of this book is shown in Figure 0.1. The backbone contains the main sections on the principles of the mathematical modelling of continuous media. Potentially these could be studied on their own; however, the resultant course would be colourless. Branching from the backbone, to the right on the diagram, are shown the sections where concepts are applied and developed in a variety of physical situations. A selection of these topics may be tailored to the main thrust of the course, as long as interposing sections are also covered. Generally, the applications furthest to the left are the simplest to cover.

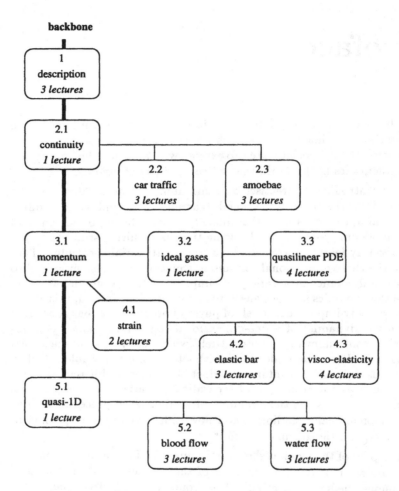

Figure 0.1: the structure of the book is primarily a backbone of modelling concepts with side branches on how these may be applied to describe a variety of physical dynamics. Very roughly, the difficulty of the sections increases from left to right, while the sophistication increases from top to bottom.

A number of exercises are included at the end of each section. These have been chosen: to provide practise in the topics discussed; to fill in the exploration of the dynamics discussed in the text; and to point the way to extensions of the work. An indication of the level of difficulty is given next to each problem.

There is a sixth chapter on miscellaneous further applications of one-dimensional continuum mechanics. These give a flavour of the wide variety of one-dimensional models. Just the essence of the modelling is discussed in these sections. Nonetheless, they show a little of how one-dimensional models are pervasive in our understanding of the universe.

This book has been written with the help and encouragement of many people. I especially thank Prof. Ernie Tuck for his many suggestions and help with the manuscript. Bill Young and Ren Potts have also provided much inspiration. Lastly I thank Barbara for putting up with me while I buried myself for weeks at a time.

Contents

Chapter 1

Describing the motion of a continuum

For a system with a small number of elements it is possible to use elementary dynamics (Newton's Laws for example) to make predictions in any given circumstance.

But what can be done for a system with a large number of elements? For example, when interested in the dynamics of air we would find that just one litre of air contains roughly 10^{23} air molecules, while the fastest computers to date do up to 10^{10} operations *per second*, thus taking roughly 10^{13} seconds or about $100,000$ *years* just to reference each molecule once. Thus, the prediction of the behaviour of air in almost any practical situation is completely impossible by molecular dynamics. The methods of continuum mechanics gives a possible route.

The fundamental **continuum hypothesis** is that the behaviour of many physical systems (in particular we shall consider materials which are gases, liquids or solids, but not exclusively) is essentially the same as if they were perfectly continuous. Everyday experience supports this hypothesis, at least for air, water, steel, etc.

Physical quantities, such as mass and momentum, associated with the molecules or elements contained within a given volume are regarded as being spread uniformly over the volume instead of being concentrated in each molecule or element.

In this chapter we use this idea to quantify and describe the movement of material making up a one-dimensional continuum. Our concerns are with fundamental dependent quantities such as density ρ, velocity v and flux q. The coordinate system which we shall use is simply the x-axis, as shown in Figure 1.1, on which the molecules or elements of the continuum move as time t varies.

Figure 1.1: The coordinate system for a one-dimensional continuum with an example distribution of molecules (shown as discs)

1.1 Averaging

Suppose we wish to measure the density of some material at some point x and at some time t. We may do this by considering a length L of the continuum, centred at x, and measuring the mass m of material in this interval; then the density at x is approximately given by $\rho = m/L$. Of course the result is different for different sized intervals, and for different positions x and times t; hence, to show all the dependencies we write

This is precisely how the density would be measured in experiments!

$$\rho(L; x, t) = \frac{m(L; x, t)}{L} \ . \tag{1.1}$$

For a fixed position at a fixed time the result for different sized intervals is qualitatively as shown in Figure 1.2. The characteristic feature of such a graph is the presence of three distinct regions: in region I the average density $\rho(L)$ is dominated by microscopic molecular fluctutations, caused by the influence of chance when there are only a small number of molecules in an averaging length; in region II $\rho(L)$ is essentially constant; while in region III $\rho(L)$ varies smoothly, the variations being caused by the macroscopic non-uniformity of the material.

Example 1.1 The molecules (mass m) of a continuum are located so that molecule i is at $x_i = \frac{i}{1+i/1000}$ for $i = \dots, -2, -1, 0, 1, 2, 3, \dots$ (note that the particles are labelled in increasing order in this example).

(a) Clearly the molecule at $x = 0$ is labelled $i = 0$, but what is the label of the molecule at $x = 500/3$? Now

$$x = \frac{i}{1 + \frac{i}{1000}} \quad \Leftrightarrow \quad i = \frac{x}{1 - \frac{x}{1000}} = \frac{500/3}{1 - \frac{500/3}{1000}} = 200$$

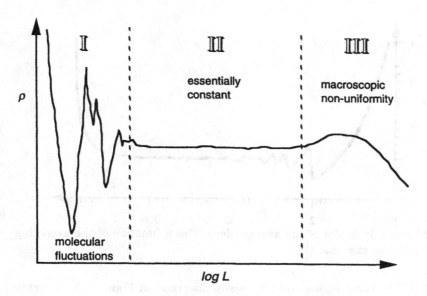

Figure 1.2: An example of the estimated density of a continuum as the averaging length L is varied from very small to vary large

Thus we can note that there are 200 molecules in the interval $(0, 500/3]$ of the x-axis.

(b) We now find the *average density* on an interval of length L (the averaging length) centred at the origin (assuming that this is a point of interest); take the interval to be $(-L/2, L/2]$.

1. The molecule at or immediately to the left of a position x is $i = \left[\frac{x}{1-x/1000}\right]$ where [ ] denotes the integer part function.

2. The molecule just at or to the left of $\frac{L}{2}$ is $i = \left[\frac{L}{2-L/1000}\right]$.

3. The molecule just at or to the left of $-\frac{L}{2}$ is $j = \left[\frac{-L}{2+L/1000}\right]$.

4. The number of particles in $\left(-\frac{L}{2}, \frac{L}{2}\right]$ is $N = i - j$.

Thus the average density at $x = 0$, as a function of L, is

$$\rho(L) = \frac{Nm}{L} = \frac{(i-j)M}{L} = \frac{m}{L}\left\{\left[\frac{L}{2-L/1000}\right] - \left[\frac{-L}{2+L/1000}\right]\right\},$$

and is plotted in Figure 1.3.

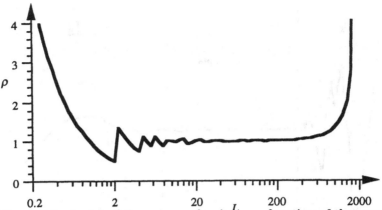

Figure 1.3: A plot of the average density as a function of the averaging length for example 1.1.

(c) The three regions may be easily discerned in Figure 1.3. Roughly speaking: region I is $L < 20$; region II is $20 < L < 500$; and region III is $L > 500$.

The **density of the continuum** at x and t is assigned the value of the average $\rho(L; x, t)$ for averaging lengths in region II (as if we extrapolated the data in region II to zero length). Note that in one dimension the density has units of *mass per length*, rather than the units *mass per volume* suitable for three-dimensional mechanics.

ASIDE   There is a form of *uncertainty principle* which is relevant here. To calculate a density we have to average over some length L, which means that we are a little uncertain about the precise location of this density; its location being uncertain by a distance of order L. The calculated density is also uncertain by some amount due to the molecular fluctuations; this uncertainty can only be decreased by increasing L. Thus the more precisely we try to find the density the less certain we are about the location of that density.

In continuum mechanics a **mathematical point** is thus physically a length big enough to smooth molecular fluctuations but is much smaller than the macroscopic variations (i.e. in region II). The accuracy of the continuum hypothesis is directly dependent upon the separation of the length scale of molecular fluctuations and the length scale of macroscopic variations, i.e. the extent of region II. In solids, liquids and gases regions I and III are separated by at least a factor of 10^5 in linear dimension (10^{15} in volume).

Note that similar graphs may be drawn for *all* physical quantities, not just the density. Thus to each point in space-time we can assign a value for the density, or velocity, or acceleration, *etc.* which is calculated by this averaging process. This averaging then defines *scalar fields* which we investigate to make predictions about the behaviour of the continuum.

Exercises

Problem 1.1 A one-dimensional continuum is composed entirely of EASY particles of unit mass. At one particular instant the ith particle happens to be located at $x_i = i$ and has a velocity of $v_i = 2 + 2 \times 10^{-3} i + 10^{-6} i^2$, for $i = \ldots, -1, 0, 1, 2, \ldots$.

(a) Find the average density $\rho(L)$ of the continuum in the interval $[-\frac{L}{2}, \frac{L}{2}]$ in terms of $n = [\frac{L}{2}]$ = the largest integer $\leq \frac{L}{2}$. Plot $\rho(L)$ versus $\log_{10} L$ for L ranging from 0.3 to 3×10^3.

(b) Similarly plot the average momentum density $p(L)$ of the continuum in the interval $[-\frac{L}{2}, \frac{L}{2}]$ versus $log_{10} L$ for L from 0.3 to 3×10^3. Hint: $\sum_{i=1}^{n} i = n(n+1)/2$ and $\sum_{i=1}^{n} i^2 = n(n+1)(2n+1)/6$.

Problem 1.2 Assume that the probability P of exactly one car being MEDIUM located in any fixed *short* segment of a one-lane highway, length δx, is proportional to the length, namely $P = \lambda \delta x$. Also assume that the probability of two or more cars in that short segment of highway is negligible.

(a) Show that the probability of there being precisely n cars along a highway of length x, $P_n(x)$, satisfies the **Poisson distribution**, $P_n(x) = (\lambda x)^n e^{-\lambda x}/n!$. (*Hint:* Consider $P_n(x + \delta x)$, in terms of the probabilities of n and $n-1$ cars being in a length x and none or one of the cars being in a length δx, and then form a differential equation for $P_n(x)$ in terms of $P_{n-1}(x)$)

(b) Given the above expression for $P_n(x)$, evaluate and interpret the following quantities: (i) $P_n(0)$, $n > 0$; (ii) $P_0(L)$, $L > 0$; (iii) $P_1(L)$, $L > 0$.

(c) Calculate the expected number of cars on a highway of length L, and also calculate the variance. Hence find the average density of cars, and find the typical size of the fluctuations about this density if an averaging length L is used to estimate it. Interpret these results in the context of attempting to model car traffic as a one-dimensional continuum.

1.2 Lagrangian description

We now turn to how to describe the motion of a continuum using averaged quantities. The term **particle** refers to a very small section of the continuum material, such as a drop of water, a few grains of sand, or a group of cars. By contrast, terms such as *molecules* or *elements* will refer to those atomic (indivisible) lumps which make up the material of the continuum. In a mathematical model a particle is vanishingly small.

Just as particles in ordinary mechanics are labelled by a countable subscript, we label particles of the continuum so that we know what each part of the material is doing. A enormous variety of labelling schemes are possible; generally the most useful is as follows. At some reference time, usually $t = 0$, each particle will have a certain position; choose the coordinate of that position to label the particle. Thus particle ξ refers to the particle which was at $x = \xi$ at time $t = 0$. The scalar functions such as $\rho = \rho^L(\xi, t)$ and $v = v^L(\xi, t)$ form part of a **Lagrangian description** of the movement of the continuum; the superscript ^{L} is used to denote a Lagrangian function.

Like house numbering along a street

Actually invented by Euler.

In particular, the function $x^L(\xi, t)$ gives the current position of the particle which was at $x = \xi$ at time $t = 0$. Consequently, just as for ordinary mechanics, the velocity v and acceleration a of the continuum particles are given by

$$v^L(\xi, t) = \frac{\partial x^L}{\partial t} \quad \text{and} \quad a^L(\xi, t) = \frac{\partial v^L}{\partial t} = \frac{\partial^2 x^L}{\partial t^2} \ . \qquad (1.2)$$

Example 1.2 Suppose particles of a continuum move according to $x = x^L(\xi, t) = \xi + \xi t^2$. Observe that at $t = 0$, $x^L(\xi, 0) = \xi$ so the Lagrangian labelling is correct. The path of each particle of the continuum (consider ξ to be fixed) is a parabola as t varies, as shown in the space-time diagram Figure 1.4. Observe that as time progresses the particles spread out; we could imagine this deformation being that of a rod or spring which is being stretched. To investigate more aspects of the motion we calculate

$$v^L(\xi, t) = \frac{\partial x^L}{\partial t} = 2\xi t \quad \text{and} \quad a^L(\xi, t) = \frac{\partial v^L}{\partial t} = 2\xi \ .$$

It is clear that the particles have constant acceleration, but the acceleration is different for each particle.

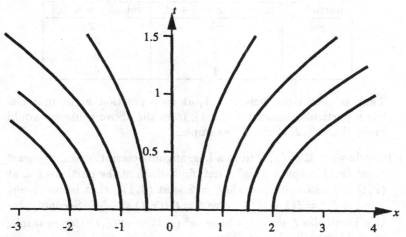

Figure 1.4: Particle paths of the continuum movement used in Example 1.2.

1.3 Eulerian description

Instead of describing quantities in the continuum as functions of the particle label ξ, an **Eulerian description** describes quantities as functions of position x (and time t). For example, asserting that $v = v^E(x,t)$ (the superscript ^{E} referring to an Eulerian description) says that the particle which is currently at position x has velocity $v^E(x,t)$. The usual system to use is the Eulerian description.

Also invented by Euler

A quantity of temporary importance is $\xi^E(x,t)$ which gives the initial position (at time $t = 0$), or equivalently the label, of the particle which at time t is at position x. It is the inverse function of $x = x^L(\xi,t)$. For example, if $x = x^L(\xi,t) = \xi + Ut$, describing the translation of a continuum with constant velocity U, then the initial position of the particle currently at x is $\xi = \xi^E(x,t) = x - Ut$.

Example 1.3 Consider further the deformation of Example 1.2.

(a) At any specific time, say $t = 1$, the following table exists:

particle ξ	location $x = \xi + \xi t^2$	velocity $v = 2\xi t$
-1	-2	-2
0	0	0
1	2	2
2	4	4

Thus at each time, here $t = 1$, at each position x the material has a particular velocity $v^E(x, 1)$; from the above table we would know that $v^E(2, 1) = 2$ for example.

(b) How do we find $v^E(x, t)$ from a Lagrangian description for a general point (x, t) in space-time? First, find which of the particles ξ is at (x, t) (for example, particle $\xi = 2$ is at $(4, 1)$), that is, we invert $x = \xi + \xi t^2 = \xi(1 + t^2)$ to give $\xi = \xi^E(x, t) = \frac{x}{1+t^2}$. Secondly, for this particular ξ we must have $v^E(x, t) = v^L(\xi, t)$ (for example, $v^E(4, 1) = v^L(2, 1) = 4$). Thus

$$v^E(x, t) = v^L\left(\xi^E(x, t), t\right) = 2\left(\frac{x}{1+t^2}\right) t = \frac{2xt}{1+t^2} .$$

(c) Similarly for the acceleration

$$a^E(x, t) = a^L\left(\xi^E(x, t), t\right) = 2\left(\frac{x}{1+t^2}\right) = \frac{2x}{1+t^2} .$$

1.4   The material derivative

An ant moves continuously; a flea jumps

Consider a rod which is being heated at one end. An ant sitting still on the rod will experience an increasing temperature. If the ant starts to move towards towards the heated end then it will experience a more rapidly increasing temperature; however, if it moves away from the heated end then the ant's increase in temperature will be slowed, the temperature it experiences may even decrease. Thus the ant's experience of temperature depends upon whether or not it is moving, and how fast it is moving. Because the particles in a continuum are typically moving in a deformation, this difference between what a stationary observer (sitting ant) experiences and what a moving particle (moving ant) experiences is fundamental in continuum mechanics.

The relationship between an Eulerian description (which is that of a fixed observer at position x) and a Lagrangian description (which is that of an observer attached to a moving particle ξ) quantifies this difference. If f is any measurable quantity or property of the material (temperature or velocity for example) then use $f^L(\xi, t)$ to denote its Lagrangian description and use $f^E(x, t)$ to denote its Eulerian description. Since these two functions describe the same scalar field, we must have $f^E(x, t) = f^L(\xi, t)$ provided that the arguments x and ξ correspond to each other, as already discussed in Example 1.3. Thus we may write

$$f^E(x, t) = f^L(\xi^E(x, t), t) \ , \tag{1.3}$$

i.e. the quantity f at (x, t) is the same as that associated with the particle ξ which is currently at the position. It is also true that

$$f^L(\xi, t) = f^E(x^L(\xi, t), t) \ , \tag{1.4}$$

i.e. the quantity f as "seen" by a particle ξ is the same as that found at the location x of the particle.

Consequently, differentiating this last equation (1.4) with respect to time and using the chain rule, we observe

$$\frac{\partial f^L}{\partial t} = \frac{\partial f^E}{\partial t} + \frac{\partial f^E}{\partial x}\frac{\partial x^L}{\partial t} = \frac{\partial f^E}{\partial t} + v^L \frac{\partial f^E}{\partial x} = \frac{\partial f^E}{\partial t} + v^E \frac{\partial f^E}{\partial x} \ ,$$

as v^L is the same as v^E provided their arguments correspond. This equation is very important: it gives an expression (the right-hand-side) *in the Eulerian system* for the rate of change of any quantity *as observed by a particle* of the continuum. This expression is frequently denoted by

$$\frac{Df^E}{Dt} = \frac{\partial f^E}{\partial t} + v^E \frac{\partial f^E}{\partial x} \ , \tag{1.5}$$

and is called the **material derivative**.

The material derivative has two components: $\frac{\partial f^E}{\partial t}$, the rate of change of f at a fixed position; and $v^E \frac{\partial f^E}{\partial x}$, the rate of change of f which a particles "sees" because it is moving through a spatially non-uniform field of f.

Example 1.4 In an Eulerian description, the acceleration of the particle at position x at time t is *not* $\frac{\partial v^E}{\partial t}$, instead it is

$$a^E = \frac{Dv^E}{Dt} = \frac{\partial v^E}{\partial t} + v^E \frac{\partial v^E}{\partial x} \ ,$$

as the acceleration is a property of the particles of the continuum. For example, to calculate the Eulerian acceleration of the deformation given in Example 1.3, without using the Lagrangian description, use the material derivative to find

$$a^E = \frac{Dv^E}{Dt} = \left(\frac{2x}{1+t^2} - \frac{4xt^2}{(1+t^2)^2} \right) + \frac{2xt}{1+t^2} \frac{2t}{1+t^2} = \frac{2x}{1+t^2},$$

as before!

We normally use an Eulerian description

Henceforth, we shall use an Eulerian description by default (unless otherwise stated). The superscripts ^{E} and ^{L} will be omitted unless they are needed to avoid confusion.

Exercises

EASY

Problem 1.3 A one-dimensional continuum is deforming according to the Lagrangian description $x = x^L(\xi, t) = \frac{\xi}{1+\xi t}$ for $\xi \geq 0$.

(a) Plot in the xt-plane the location of the material particle which was initially at the location $x = 1$ at $t = 0$.

(b) Find the Lagrangian descriptions of the velocity, $v^L(\xi, t)$, and acceleration, $a^L(\xi, t)$.

(c) Hence find the Eulerian description of the velocity, $v^E(x, t)$.

(d) Verify that the material derivative of the Eulerian velocity, $\frac{Dv^E}{Dt}$, is the same as the acceleration of the particles of the material, a^L.

MEDIUM

Problem 1.4 This type of problem is precisely that which is solved when predicting the dispersal of pollutants, e.g. the radioactive material released from the accident at Chernobyl in 1986.

(a) Suppose that a deformation of a continuum is described by $v^E(x, t)$, some known function, and that the paths of the particles in the continuum need to be found. Explain why $x^L(\xi, t)$ satisfies the first-order differential equation (in terms of the known function v^E) $\frac{\partial x^L}{\partial t} = v^E(x^L, t)$.

(b) Given that $v^E(x, t) = \frac{2tx}{1+t^2} + 1 + t^2$, solve the differential equation obtained in part (a) to find $x^L(\xi, t)$. Suppose that at time $t = 0$ the particles in the interval $[0, 1]$ of the continuum are contaminated with some pollutant; where will these particles be at time $t = 2$?

(c) For the deformation given in part (b) find ξ^E, v^L and a^L. Verify that the acceleration of particles of the continuum, $a^L(\xi, t)$, is also given by $\frac{Dv^E}{Dt}$.

Problem 1.5 *Note that v^E is not given by $\frac{\partial \xi^E}{\partial t}$.* Use the nature of the EASY material derivative and the nature of $\xi^E(x, t)$ to explain why $\frac{D\xi^E}{Dt} = 0$. Hence deduce that $v^E(x, t) = -\frac{\partial \xi^E/\partial t}{\partial \xi^E/\partial x}$. Confirm this result for the deformations given in Problems 1.3–1.4.

Chapter 2

Conservation of material

Common to all continuous media is the principle that matter can be neither created nor destroyed. In continuum mechanics this is a non-trivial statement and leads to the derivation of the vitally important *continuity equation*. It applies to the evolution of *every* continuum as it just involves the density and the movement of the material. With this equation we may then immediately investigate some applications of continuum mechanics.

2.1 The continuity equation

2.1.1 Slicing

To make progress we need an additional tool which enables us to translate physical principles into mathematical equations. *Conceptually*, we extract and investigate a typical length or interval of the continuum, looking at the physical processes which take place both in it and on it. The slice of the continuum so examined must be big enough so that molecular fluctuations are totally irrelevant. Having quantified the physical processes, we then use the following theorem to derive a corresponding governing equation.

Theorem : *If $f(x)$ is continuous on $[c, d]$ and $\int_a^b f(x)\,dx = 0$ for all a, b such that $c < a < b < d$ then $f(x) = 0$ for all x in $[c, d]$*

The only theorem in this book

13

Proof : (by contradiction)

Assume the theorem is false

$\Rightarrow \exists x_0 \in (c, d)$ such that $f(x_0) \neq 0$, say $f(x_0) > 0$ (the other case of $f(x_0) < 0$ is nearly identical)

$\Rightarrow$ (from continuity) $\exists \delta > 0$ such that $f(x) > 0$ for all x in $(x_0 - \delta, x_0 + \delta)$

$\Rightarrow$ upon letting $a = x_0 - \delta$ and $b = x_0 + \delta$, that $\int_a^b f(x)\,dx > 0$ since $f(x) > 0$ on $[a, b]$ which is a contradiction!

Thus the theorem is true (that $f(c) = f(d) = 0$ follows from the continuity of f)

ASIDE

> There are other ways of proceeding from physical principles to mathematical equations, but they all rely on the same physical arguments about what is going on in an interval of the continuum. One popular way uses a "small" length of the continuum and expands quantities in Taylor series' which then give rise to the governing differential equations. Another route is to use an interval which is moving with the material of the continuum (a "Lagrangian interval"). The interested reader is referred to Hodge[3, §2-2] where these methods are all used to derive precisely the same governing equation. We prefer the route provided by the above theorem because it is a clear way and may be easily modified to allow discontinuous solutions (*shocks*), which are frequently necessary.

2.1.2   Conservation of mass

Consider *any* fixed slice of a continuum, say the interval $[a, b]$ as shown in Figure 2.1. The total mass of material in $[a, b]$ is simply the integral

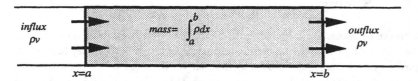

Figure 2.1: A slice of a continuum to investigate the conservation of mass

The physical processes in this slice.

of the density, namely $\int_a^b \rho\,dx$. Because material cannot appear or disappear, this total mass can only change by material entering or leaving the interval across the ends $x = a$ and $x = b$. It is the movement of the continuum which contributes to this change. If the material is moving with (Eulerian) velocity $v(x, t)$ then the **mass flux**, defined to be the rate at which material is carried to the right past a given point, is simply

ρv. Thus we may deduce

conservation of mass $\Rightarrow$ (rate of mass increase in the interval)

$$= \text{(rate of mass influx across the ends)}$$

$$\Rightarrow \quad \frac{\partial}{\partial t} \int_a^b \rho \, dx = \rho(a, t)v(a, t) - \rho(b, t)v(b, t)$$

$$\Rightarrow \quad \int_a^b \frac{\partial \rho}{\partial t} dx + \rho(b, t)v(b, t) - \rho(a, t)v(a, t) = 0$$

$$\Rightarrow \quad \int_a^b \frac{\partial \rho}{\partial t} dx + [\rho v]_a^b = 0$$

$$\Rightarrow \quad \int_a^b \frac{\partial \rho}{\partial t} dx + \int_a^b \frac{\partial (\rho v)}{\partial x} dx = 0$$

$$\Rightarrow \quad \int_a^b \left[\frac{\partial \rho}{\partial t} + \frac{\partial}{\partial x}(\rho v) \right] dx = 0$$

Since this identity is true for all $a < b$ in the continuum material, then by the previous theorem the integrand must be zero for all x. Thus the following **continuity equation**

$$\frac{\partial \rho}{\partial t} + \frac{\partial}{\partial x}(\rho v) = 0 \; , \tag{2.1}$$

must hold for all points in the continuum.

An alternative form of this equation makes it very easy to understand. Expanding the spatial derivative, the equation becomes $\frac{\partial \rho}{\partial t} + v \frac{\partial \rho}{\partial x} + \rho \frac{\partial v}{\partial x} = 0$. The interesting thing here is that the first two terms on the left-hand side are just the material derivative of the density, and thus $\frac{D\rho}{Dt} + \rho \frac{\partial v}{\partial x} = 0$, which may be rearranged to give

$$\frac{1}{\rho} \frac{D\rho}{Dt} = -\frac{\partial v}{\partial x} \; .$$

That is, the proportional rate of change of density as seen by a particle of the material, $\frac{1}{\rho} \frac{D\rho}{Dt}$ is equal to $-\frac{\partial v}{\partial x}$. This is reasonable as, if $\frac{\partial v}{\partial x} < 0$ then the particles in front are moving slower then the particles behind and so the particles must be bunching together, i.e. the density must be increasing.

Exercises

Problem 2.1 Suppose that a one-dimensional continuum has a density   EASY

$\rho(x, t)$, that mass is being transported to the right past any particular point at a rate $q(x, t)$ (due to any cause), and that mass is being continuously added to the continuum by some external agency at a rate $r(x, t)$ (if r is negative then mass is being taken from the continuum). Derive the continuity equation

$$\frac{\partial \rho}{\partial t} + \frac{\partial q}{\partial x} = r \tag{2.2}$$

for this continuum.

2.2   Car traffic

With just the continuity equation we now turn to our first application of continuum mechanics: that of car traffic. Consider the simplest situation of a stream of cars travelling along a one-lane highway, and discover the sort of effects and phenomena which continuum mechanics can describe. It is a very useful example of its application as we are all familiar with car traffic and how it behaves, and we can relate this to the predictions. Car traffic illustrates very nicely the range and the limitations of continuum mechanics.

2.2.1   A continuum?

Unlike gases, liquids or solids

Car traffic, as illustrated in Figure 2.2, is very obviously molecular, or discrete, with the cars being the molecules. However, this is merely a reflection of our human scale of perception—if we were flying high above the road then we would no longer be able to see each individual car, we would only be able to discern places where there were a lot of cars and places where there were few. It is on this large scale that we can treat car traffic as a continuum.

To describe car traffic as a continuum we need to average over a length of highway (or over a time interval) which includes a reasonable number of cars, yet which is smaller than the large-scale density and velocity variations which we wish to describe. Say we want something like 20 cars to average over, then we would need an averaging length of about 100 metres in heavy traffic or about 1 km in light traffic, a corresponding averaging time would be about one minute. The continuum hypothesis for car traffic is clearly best in heavy traffic or over long stretches of road; even so, it is only a rough approximation. Nonetheless, the qualitative results are remarkably impressive.

Figure 2.2: A picture of a one-lane highway with a stream of cars travelling on it.

2.2.2 The density-velocity relation

So far we have two unknowns: the density of the cars ρ, measured in *cars per km*; and the velocity of the cars v, measured in *km per hour* or sometimes in *km per minute*. But we have only one equation, namely the continuity equation $\frac{\partial \rho}{\partial t} + \frac{\partial}{\partial x}(\rho v) = 0$. Now, when solving a set of linear equations, (usually) we need as many equations as unknowns in order to have a well-posed problem. The same principle holds for this more complicated situation and so we need another equation to "close the problem".

This extra equation takes the form of an experimentally determined "equation of state". We argue that in light traffic, car drivers drive at the speed limit (more-or-less!) while in heavy traffic they slow down due to the close proximity of the neighbouring cars and the likelihood of a crash if they all travelled fast. A reasonable model of this behaviour is to pose

$$v = V(\rho) , \tag{2.3}$$

where $V(\rho)$ would look as plotted in Figure 2.3. This relation just says that we expect car drivers to drive at a velocity which depends only upon the density of cars in their own vicinity: travelling at the speed limit $v = 60$ km/hr at low densities; slowing down at high densities; and stopping $v = 0$ if the cars are bumper-to-bumper at a density of $\rho_j \approx 150$ cars/km. Experimental data [2, p288] supports this simple relation between car velocity and car density.   *j for traffic jam.*

It is typical of continuum mechanics that experiment is needed to close the problem.

In theory it is possible to proceed from a description of the atomic or molecular interactions in a material to a strict description of the continuum behaviour *without* any extra experimental information. While this   ASIDE

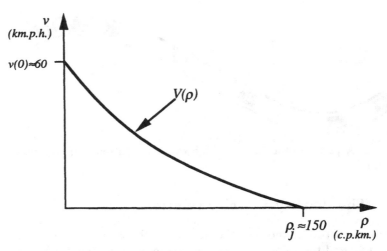

Figure 2.3: A typical relation found between car velocity v and car density ρ.

has been done fairly simply for car traffic, based on so-called car-following models [2, §64], to do this for gases, liquids and solids has been a research problem for the last 100 *years*; it remains a research problem to this day, even for the simplest of gases. One of the interesting unresolved issues being that no-one is sure how the reversibility of particle mechanics is lost when transforming to a continuum description.

Since the velocity v only appears in the combination ρv in the continuity equation, a more useful quantity to determine by experiment is the flux of cars q (the number of cars going past a given point in an interval of time) as a function of the density ρ:

It is easier to observe $Q(\rho)$ than $V(\rho)$.

$$q = Q(\rho) = \rho V(\rho) \ .$$

Given the typical density-velocity relation shown in Figure 2.3, a typical car flux-density relation is shown in Figure 2.4.

One important feature to observe in such a flux–density relation is the presence of a maximum in the flux, at $q_m \approx 1500$ cars/hr $= 25$ cars/min. In terms of maximising the effectiveness of a road, from the viewpoint of the system as a whole, this maximum is the point at which the roads should operate. It occurs at a density $\rho_m \approx 50$ cars/km and corresponds to a surprisingly low velocity of $v_m \approx 30$ km/hr.

m for maximum flux

But not from the viewpoint of an individual.

Example 2.1 Experiments in the Lincoln Tunnel, New York found that a good fit to the data (except at very low densi-

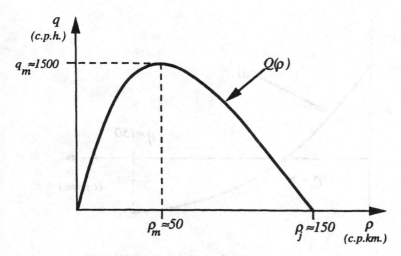

Figure 2.4: A typical relation found between the car flux $q = \rho v$ and the car density ρ.

ties) is obtained by $Q(\rho) = v_m \rho \log(\rho_j/\rho)$ where the two constants were found to be $v_m = 27.5$ km/hr and $\rho_j = 142$ cars/km. Differentiating $Q(\rho)$ to find the location of the maximum gives $Q'(\rho) = v_m [\log(\rho_j/\rho) - 1]$, from which we readily deduce that the maximum occurs at a density of $\rho = \rho_j/e = 52.2$ cars/km and corresponds to a flux of cars of $q_m = 1436$ cars/hr, and all travelling at a velocity of v_m.

Returning to the continuity equation, we use $\rho v = Q(\rho)$ to eliminate the unknown v and obtain

$$\frac{\partial \rho}{\partial t} + \frac{\partial}{\partial x}Q(\rho) = 0 , \quad \text{that is} \quad \frac{\partial \rho}{\partial t} + Q'(\rho)\frac{\partial \rho}{\partial x} = 0 .$$

We use this last form of the continuity equation, and so note that what we are really interested in is $Q'(\rho)$. Letting $c(\rho) = Q'(\rho)$, car traffic must evolve so that

$$\frac{\partial \rho}{\partial t} + c(\rho)\frac{\partial \rho}{\partial x} = 0 , \tag{2.4}$$

a single non-linear hyperbolic differential equation for $\rho(x,t)$, where $c(\rho)$ typically looks as plotted in Figure 2.5.

Equation (2.4), together with velocity-density relation (2.3), forms the mathematical model of traffic flow. Some of the predictions which

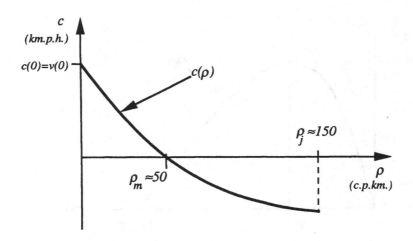

Figure 2.5: A typical plot of $c(\rho) = Q'(\rho)$

may be made from such a model are discussed in the following sub-sections. In particular, $c(\rho)$ is the velocity at which fluctuations in traffic density progress—c is, in other words, the speed of waves in the car traffic continuum.

2.2.3 Nearly uniform traffic flow

There is a simple exact solution of the governing equation (2.4); that of constant density

$$\rho(x,t) = \rho_* \ , \tag{2.5}$$

where ρ_* is any constant. In this *uniform state*, or *equilibrium* state, the cars are equi-spaced along the highway and they all travel with the same velocity, namely $V(\rho_*)$. While this uniform state may indeed occur in some situations, it is of little interest as there is no evolution in time.

Some illustrative *approximate* solutions of traffic flow may be found by solving the **linearised** equations. Suppose that the traffic density does vary, but only varies a "little" from some constant reference density ρ_*. That is, suppose

$$\rho = \rho_* + \hat{\rho}(x,t) \ ,$$

where ρ_* is constant and the variations $\hat{\rho}$ are deemed "small" in some sense. The meaning of **small** is that we must consistently neglect products of "small" terms. For example, $c(\rho)$ occurs in equation (2.4) and we rewrite this, using a Taylor's series in $\hat{\rho}$, as

$$c(\rho) = c(\rho_* + \hat{\rho}) \;=\; c(\rho_*) + c'(\rho_*)\hat{\rho} + \frac{1}{2}c''(\rho_*)\hat{\rho}^2 + \cdots$$
$$\approx\; c(\rho_*) + c'(\rho_*)\hat{\rho} \;,$$

as all the higher powers of $\hat{\rho}$ are negligible due to their smallness. The governing differential equation (2.4) then becomes

$$\frac{\partial}{\partial t}(\rho_* + \hat{\rho}) + [c(\rho_*) + c'(\rho_*)\hat{\rho}]\frac{\partial}{\partial x}(\rho_* + \hat{\rho}) = 0$$

$$\Rightarrow \quad \frac{\partial \hat{\rho}}{\partial t} + c(\rho_*)\frac{\partial \hat{\rho}}{\partial x} + c'(\rho_*)\hat{\rho}\frac{\partial \hat{\rho}}{\partial x} = 0 \;.$$

However, the third term on the left-hand-side of the above equation is a product of "small" terms and must also be neglected in this approximation. Thus we derive that small density fluctuations in traffic density obey

$$\frac{\partial \hat{\rho}}{\partial t} + c_*\frac{\partial \hat{\rho}}{\partial x} = 0 \;, \tag{2.6}$$

where $c_* = c(\rho_*)$ is a constant. This is often called the **unidirectional wave equation**.

This differential equation has a very simple general solution, namely D'Alembert's solution $\hat{\rho} = f(x - c_* t)$ where f may be *any* differentiable function and is given by the initial condition of the car traffic. This may be readily verified by simply substituting it into (2.6); upon doing this we find

$$\frac{\partial \hat{\rho}}{\partial t} + c_*\frac{\partial \hat{\rho}}{\partial x} = f'(x - c_* t)(-c_*) + c_* f'(x - c_* t) = 0 \;.$$

In any given circumstance the function f is determined by the car density at some initial time, say $t = 0$. Suppose that the density $\rho = \rho_0(x)$ at time $t = 0$ where $\rho_0(x)$ is some known function of x. But D'Alembert's solution shows that $\rho_0(x) = \rho(x, 0) = \rho_* + \hat{\rho}(x, 0) = \rho_* + f(x - c_* 0) = \rho_* + f(x)$. Hence we find that $f(x) = \rho_0(x) - \rho_*$. Thus finally, car traffic whose initial density $\rho_0(x)$ is everywhere close to ρ_* evolves approximately like

$$\rho(x, t) = \rho_* + f(x - c_* t) = \rho_0(x - c_* t) \;. \tag{2.7}$$

This solution shows that the density at (x, t) is the same as the initial density at location $x - c_* t$ which is a distance $c_* t$ to the left of

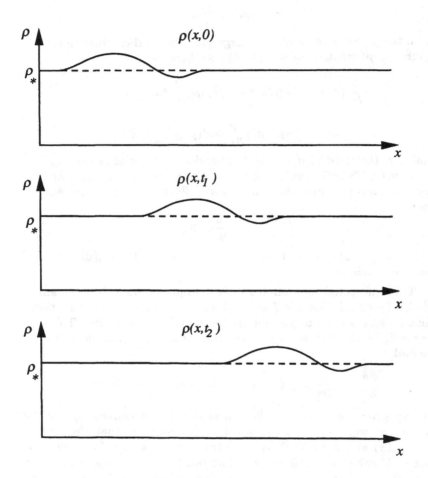

Figure 2.6: An example of the approximate car traffic solution, Equation (2.7).

x. Thus the shape of the car density remains constant in time, it just translates at the constant velocity c_* without change in form, as shown in Figure 2.6. Another way to view this solution is to say that the density ρ is constant along lines $x - c_* t = \text{const}$. Thus we may draw a diagram like Figure 2.7 which shows these lines, and then say that the value of the density is "carried" unchanged along these lines. This

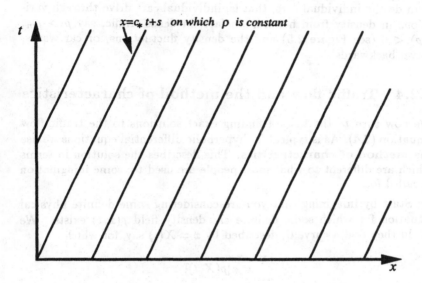

Figure 2.7: A diagram of the approximate lines in xt-space on which the density ρ is constant, and hence must take the value it has at the initial time.

turns out to be a very useful viewpoint because in the next sub-section we show that the exact solution of the car traffic equation (2.4) can be described by such a diagram—the only difference between the exact solution and the approximate solution found here is that the slopes of the lines in the approximate diagram are not quite correct.

Observe that the velocity at which the shape of the car density travels is c_* which is not the same as the velocity at which the cars themselves travel. The difference between these two velocities should be no great surprise because we are very familiar with such a phenomenon: on the surface of water we see water waves move while the water as a whole stays still; in air we hear sound because it propagates at a velocity different to that of the wind. In car traffic we see that the wave velocity

Wave velocity versus car velocity

is
$$c(\rho) = Q'(\rho) = V(\rho) + \rho V'(\rho) \ .$$

Firstly, observe $c(0) = V(0)$ so that in light traffic the density fluctuations travel at the same velocity as the cars do; that is, there is little interaction between cars. Secondly, in general $V'(\rho) < 0$ (just see Figure 2.3) and so $c(\rho) < V(\rho)$ so that the density fluctuations travel slower than do the individual cars; that is, individual cars drive through variations in density from behind. In fact, in dense traffic, say $\rho > \rho_m$, $c(\rho) < 0$ (see Figure 2.5) and the density fluctuations, or car waves, travel backwards!

2.2.4   Traffic flow and the method of characteristics

We now turn to the task of finding exact solutions to the traffic flow Equation (2.4). As is typical for hyperbolic differential equations we use the **method of characteristics**. This describes the solution in terms which are different to what most people are used to; some imagination is called for.

Start by imagining that you are considering some definite physical situation for which some definite car density field $\rho(x,t)$ exists. We could then find a curve C, described by $x = X(t)$ say, for which

Imagine solving this as a first-order D.E. for the unknown $X(t)$

$$\frac{dX}{dt} = c\left[\rho(X,t)\right] \ .$$

The curve C is called a **characteristic**. Then on C we can consider ρ to be purely a function of t, namely $\rho = \rho(X(t), t)$. Differentiating with respect to t then gives

$$
\begin{aligned}
\frac{d\rho}{dt} &= \frac{\partial \rho}{\partial t} + \frac{\partial \rho}{\partial x}\frac{dX}{dt} \quad \text{by chain rule} \\
&= \frac{\partial \rho}{\partial t} + c[\rho]\frac{\partial \rho}{\partial x} \quad \text{by definition of } C \\
&= 0 \quad \text{by the governing equation (2.4)}
\end{aligned}
$$

Thus on each such characteristic C the density ρ must be constant. Remember that this is precisely the same statement as was illustrated in Figure 2.7 for the earlier approximate solution, and there it just represented simple wave propagation.

We make considerably more progress by further deducing:

$$\rho \text{ is constant on } C$$

$$\Rightarrow \quad c(\rho) \text{ is constant on } \mathcal{C}$$
$$\Rightarrow \quad \frac{dX}{dt} \text{ is constant on } \mathcal{C}$$
$$\Rightarrow \quad \mathcal{C} \text{ is a straight line !}$$

With these observations we now draw the solution of the car traffic problem—equation (2.4) with the initial car density given as $\rho = \rho_0(x)$ at time $t = 0$, where $\rho_0(x)$ is some known function.

The procedure suggested by the above is as follows:

- at any point $x = s$ (and $t = 0$), calculate $c_0(s) = c\left[\rho_0(s)\right]$;

- then the straight line through $(x,t) = (s,0)$ and of slope $c_0(s)$, namely $x = s + c_0(s)t$, is a characteristic $\mathcal{C}$ on which ρ is constant, namely $\rho_0(s)$;

- thus a parametric solution of equation (2.4) is

$$\rho = \rho_0(s) \quad \text{on} \quad x = s + c_0(s)t$$

and so at any time t we vary s and plot $\rho_0(s)$ versus this $x(s)$ to obtain a graph of $\rho(x,t)$.

This completes the characteristic solution and we now see how it works in a couple of interesting examples.

Example 2.2 Consider a uniform stream of cars except that a group of cars are bunched closer together than the rest. This gives rise to a constant density except at the bunch where there is a region of higher density, as shown in Figure 2.8.

Now, a characteristic passes through every point on the x-axis and furthermore they have inverse-slope $\frac{dx}{dt} = c_0(x) = c\left[\rho_0(x)\right]$. *Note*: we use the term **inverse-slope** to mean $\frac{dx}{dt}$, as opposed to the *visual slope* $\frac{dt}{dx}$ of curves in the xt-plane; thus lines of lesser inverse-slope appear steeper in the xt-plane. Since $c(\rho)$ is a decreasing function of density, see Figure 2.5, the function $c_0(x) = c\left[\rho_0(x)\right]$ will look as displayed in Figure 2.9. Drawing a *selection* of the characteristics emanating from the x-axis we would draw a **characteristic diagram** as shown in Figure 2.10.

On each of the characteristics the density is constant and so at any time, $t = t_1$ say, draw the line $t = t_1$ in on the characteristic diagram and then wherever a characteristic intersects the

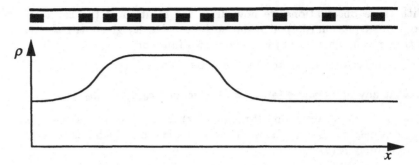

Figure 2.8: A uniform stream of cars with a localised bunching, and the associated density field $\rho_0(x)$.

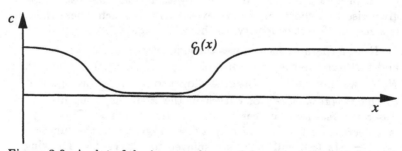

Figure 2.9: A plot of the inverse-slopes of characteristics $c_0(x)$ as a function of x at time $t = 0$ for the initial traffic density shown in Figure 2.8.

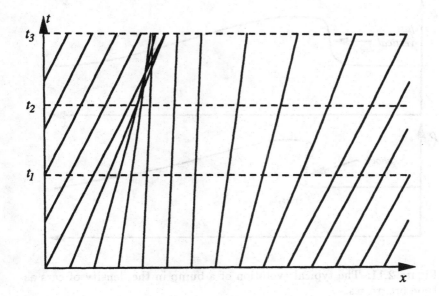

Figure 2.10: A characteristic diagram for a group of cars bunched together in an otherwise uniform stream of cars. The dashed lines are fixed times at which the density on each characteristic gives the density as a function of x.

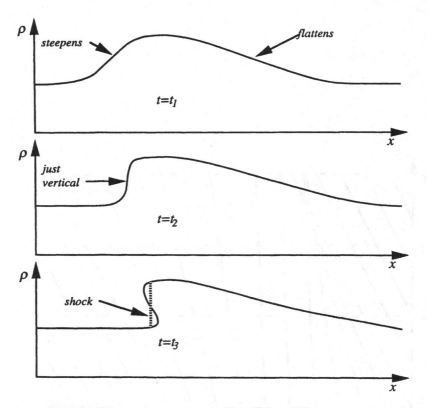

Figure 2.11: The typical evolution of a bump in the density of cars as time progresses.

$t = t_1$ line we know that the density at that particular value of x is the density associated with that characteristic from the initial instant $t = 0$. Thus from Figure 2.10 we draw the car density as time progresses; this is shown in Figure 2.11. Observe that the bump in density, *the density wave*, progresses just as in the linearised solution discussed in Section 2.2.3; the difference here is that the bump changes shape since the characteristics are not strictly parallel, as in the linearised solution. In particular, see the typical behaviour of car traffic—that the back of the bump steepens and the front of the bump flattens out, see $t = t_1$ in Figure 2.11. Since $V(\rho) > c(\rho)$, individual cars travel quicker than the density bump and thus: a car enters the bunched group

Prediction

from the back and has to brake sharply to slow down to a velocity appropriate to the higher local density; the car can only accelerate slowly through the bunch as it makes its way through the gently decreasing density at the front of the bump. This is in accord with experience.

As time increases the steepening continues until at some time, $t = t_2$ say, the back of the density bump is just vertical! For $t > t_2$ the method of characteristics predicts that the density is multivalued in some range of x, due to more than one characteristic passing through some points of the characteristic diagram 2.10! This is clearly ridiculous; the averaging process used to define a density field can only result in a unique density at each x. However, the conservation of cars must still apply, thus we may patch-up the solution by fitting a **shock** in the solution; that is, we allow $\rho(x,t)$ to jump at the right location in the multi-valued region, as indicated in the $t = t_3$ graph in Figure 2.11. We shall not pursue this any further here. The fitting of shocks can be done and can be justified, see Section 3.3.1.

wave breaking

ASIDE

Shocks occur in many other, more familiar situations: in supersonic flight a *sonic boom* is just such a shock in the gas density; in shallow water (e.g. on a beach) disturbances travel faster in deeper water and so a water wave steepens at the front until it breaks and forms a turbulent bore (running up the beach) which is just such a shock.

Example 2.3 Traffic lights An interesting situation is when there are a queue of cars waiting at a red traffic light, which then turns green. Out of interest, a numerical simulation of this is shown in Figure 2.12: for $t < 0$ there is a queue of cars waiting to the left of a red traffic light which is situated at $x = 0$; at $t = 0$ the light turns green and the lead car starts to accelerate; for $t > 0$ successive cars in the queue accelerate until they too pass the traffic light. The continuum model of car traffic predicts remarkably well all the details of the evolution of the car traffic in this situation. The appropriate initial condition on the car density is

$$\rho(x,0) = \rho_0(x) = \begin{cases} 0 & \text{if } x > 0 \\ \rho_j & \text{if } x < 0 \\ \text{all values in between} & \text{if } x = 0 \end{cases}$$

since there are no cars in front of the traffic light, and the cars are bumper-to-bumper, $\rho = \rho_j$, behind the traffic light. It is

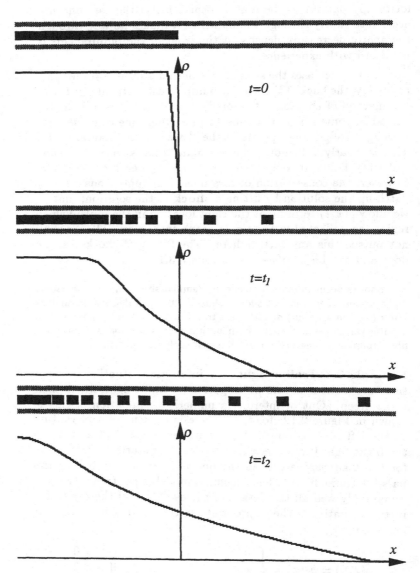

Figure 2.12: Simulation of car traffic at a traffic light which suddenly turns green. Also plotted is the resultant density field.

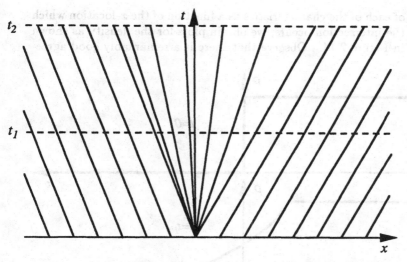

Figure 2.13: A typical characteristic diagram for car traffic which has been waiting at a red traffic light after it turns green.

convenient to consider that the density takes on all values in between 0 and ρ_j at $x = 0$ for two reasons: firstly, as discussed in Section 1.1, the density field is obtained by averaging over some length L which indeed gives a fairly smooth transition from 0 to ρ_j over a narrow region (see Figure 2.12; secondly, it is the only way to find a complete solution by the method of characteristics. The typical characteristic diagram for this situation is illustrated in Figure 2.13. There are three types of characteristic lines: the lines emanating from the region of no cars in front of the traffic light, described by $x = s + c(0)t$ for $s > 0$, carry a density of $\rho = 0$ with them; the lines emanating from the region of bumper-to-bumper cars behind the traffic light, described by $x = s + c(\rho_j)t$ for $s < 0$, carry a density of $\rho = \rho_j$ with them (note that $c(\rho_j) < 0$ and so these characteristics slope backwards); and the characteristics emanating from the traffic light are of all slopes in between, described by $x = c(\rho)t$ for $0 < \rho < \rho_j$, and carry their particular value of the density with them (depending upon their slope). These characteristics are said to form an **expansion fan**. To find the density at any particular time, say $t = t_1$, after the light turns green we just look at the intersections of the characteristics and the the line $t = t_1$ and plot the density of each of the characteristics as a function of the x-location which the intersection occurs; we obtain plots for the density as shown in Figure 2.14. Observe that there is a remarkably good agree-

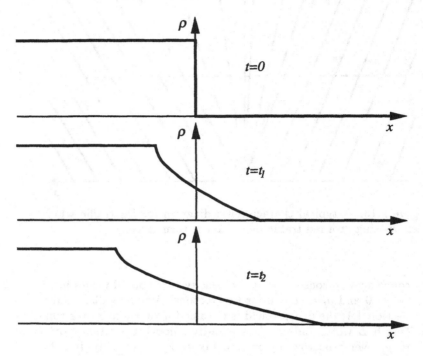

Figure 2.14: The evolution of the density of cars after a red traffic light turns green—as predicted by the method of characteristics.

Prediction

ment between these predictions and the numerical simulations. Furthermore, these both agree well with our experience of such traffic queues.

- For example, the lead car accelerates away unimpeded by any other traffic, while cars behind in the queue can only accelerate relatively slowly.

ASIDE

In the continuum model the acceleration of the leading car is instantaneous, but in reality it takes a finite time. However, the acceleration time is small compared to a reasonable averaging time interval and so it is appropriate that it be taken to be "zero" in the continuum model.

- As another example, a car at location $x = -d$ in the queue has

to wait a finite time before it starts to move (while it waits for
the surrounding density to fall below the traffic jamming $\rho = \rho_j$),
again in accord with our experience. In fact, the *wave* of cars
starting to move progresses backwards at a speed $-c(\rho_j)$ and so
the car has to wait a time $d/[-c(\rho_j)]$.

- The shape of the density variation in the expansion fan, as shown
  in Figure 2.14, is easy to describe. The characteristics carrying
  the density ρ are $x = c(\rho)t$; at any fixed t we may consider this as
  giving x as a function of ρ, namely the *inverse* function of $\rho(x)$.
  Thus the density dependence upon x is just a stretched version of
  the inverse function of $c(\rho)$.

- Lastly, consider the characteristic $x = 0$. The value of the density
  associated with this line is the one for which $c(\rho) = 0$, but $c(\rho) =$
  $Q'(\rho)$ and so this occurs at the density ρ_m corresponding to the
  maximum flux of cars q_m. While the traffic light stays green the
  car density at the traffic light is constant, but more importantly, it
  is the density of maximum car flux. Hence a properly timed traffic
  light is the most efficient way of handling an intersection, in that
  while each stream of cars have a green light, they are passing the
  intersection at the maximum number of cars per minute.

This last point about the efficiency of traffic lights was not appreci-   ASIDE
ated until this continuum model of car traffic was developed in the 1950's.
Consider this while you next wait at some lights.

2.2.5 Remarks

It is important to remember that predictions which involve changes
more rapid than the averaging length/time are not valid. However,
conservation of cars is such a *robust principle* that even though a solution
is invalid in some small region, the global picture is still correct.

A possible *source of error* is in the assumed form of the density-
velocity relation: $v = V(\rho)$. For example: we could credit drivers with
some intelligence.   Namely, if they see denser traffic ahead (in which
case $\frac{\partial \rho}{\partial x} > 0$) then they slow down; which mathematically could be
described by $v = V(\rho) - \frac{\nu}{\rho}\frac{\partial \rho}{\partial x}$, where the constant ν would be determined
by observation. This postulate and the continuity equation (2.1)

$$\Rightarrow \quad \frac{\partial \rho}{\partial t} + \frac{\partial}{\partial x}\left[\rho V - \nu \frac{\partial \rho}{\partial x}\right] = 0$$

$$\Rightarrow \quad \frac{\partial \rho}{\partial t} + Q'(\rho)\frac{\partial \rho}{\partial x} - \nu \frac{\partial^2 \rho}{\partial x^2} = 0$$

$$\Rightarrow \quad \frac{\partial \rho}{\partial t} + c(\rho)\frac{\partial \rho}{\partial x} = \nu \frac{\partial^2 \rho}{\partial x^2}$$

Compare with the heat equation

This is the same as equation (2.4) except for the new term $\nu \frac{\partial^2 \rho}{\partial x^2}$. This term is a "diffusion" term which tends to smooth out rapid variations. Thus "shocks" may start to form, but the steepening is arrested by this "look-ahead" induced diffusion.

ASIDE

If $c(\rho)$ is linear then this equation is called *Burger's equation*. Burger's equation has arisen in many branches of physics because it combines, in as simple a form as possible, the two fundamental physical processes of nonlinear steepening, represented by $c(\rho)\frac{\partial \rho}{\partial x}$, and of diffusion, represented by $\nu \frac{\partial^2 \rho}{\partial x^2}$. Curiously enough, this equation too is exactly solvable—by the nonlinear Cole-Hopf transformation!

For another example of possible errors, we could incorporate the reaction time of the drivers by postulating $v = V[\rho(x, t - \tau)]$ where τ is the reaction time. That is, the drivers travel at a velocity to suit the conditions as they existed just a little earlier in time. This postulate gives rise to a difficult class of problems, but they may be approximately solved to obtain a picture of what sort of effects the reaction time would produce, see Problem 2.10.

There are many more interesting aspects of a continuum model of traffic flow. For example: how a shock wave evolves; what happens when a traffic light turns red or goes through its cycle; finding the paths of individual cars from any given solution; what happens when cars enter from side roads; and generalisations to multi-lane highways. The book by Haberman [2] discusses many of these problems.

Exercises

EASY

Problem 2.2 Devise a *simple* experiment to measure the average density of cars on a given stretch of road (of length L, say) as a function of time (assume that no cars can enter or leave the stretch of road except at the ends). Devise a method to determine experimentally the car flux-density relation $q = Q(\rho)$.

MEDIUM
Note the change of time unit from hours to minutes

Problem 2.3 Given the car flux-density relation $Q(\rho) = \rho(1 - \rho/150)(1 - \rho/300)$ cars per minute, where ρ is measured in cars per km $(0 \le \rho \le 150)$:

(a) Sketch $Q(\rho)$, the corresponding car velocity-density relation $V(\rho)$, and the *wave speed* $c(\rho)$. Indicate the values and the corresponding points on the graphs of the maximum density of cars, the maximum flux of cars, the velocity of cars at the maximum flux, and the maximum velocity of cars.

(b) Draw accurately a graph of the characteristics for the *traffic light* problem; the initial conditions being

$$\rho(x,0) = \rho_0(x) = \begin{cases} 0 & \text{if } x > 0 \\ 150 & \text{if } x < 0 \\ \text{all values in between} & \text{if } x = 0. \end{cases}$$

Use a scale to cover the range ± 3 km and $0 - 3$ minutes.

(c) Hence graph the solution $\rho(x,t)$ at times $t = 1$, 2 and 3 minutes.

Problem 2.4   Given the car flux-density relation $Q(\rho) = \rho(1 - $ MEDIUM $\rho/150)(1 - \rho/300)$ cars per min where ρ is measured in cars per km, as in Problem 2.3.

(a) Draw the characteristics for the problem of the evolution of a *bunch of cars* for which the initial condition is:

$$\rho(x,0) = \rho_0(x) = \begin{cases} 50 - 25|x| & \text{if } |x| \leq 2 \\ 0 & \text{if } |x| \geq 2 \end{cases}$$

Use a scale to cover the range -2 km$< x < 5$ km and $0-3$ minutes.

(b) Hence, graph the predicted solution $\rho(x,t)$ at times $t = 0$, 1, 2, and 3 minutes. Discuss these predictions.

Problem 2.5   The initial value problem $\frac{\partial \rho}{\partial t} + c(\rho)\frac{\partial \rho}{\partial x} = 0$ such that EASY $\rho(x,0) = \rho_0(x)$ has the solution $\rho = \rho_0(s)$ on characteristics $x = s + c_0(s)t$ where $c_0(s) = c[\rho_0(s)]$.

(a) Regard $x = s + c_0(s)t$ as an implicit equation for the function $s(x,t)$; differentiate it to find implicit formulae for $\frac{\partial s}{\partial t}$ and $\frac{\partial s}{\partial x}$. Hence show that $\rho = \rho_0[s(x,t)]$ does satisfy the governing differential equation.

(b) The denominator $1 + c_0'(s)t$ which appears in part (a) could be zero at some positive time t. What do you imagine that this signifies?

MEDIUM   **Problem 2.6** The method of characteristics can *sometimes* look like
a Lagrangian description. Consider a radioactive cloud being carried
along by the wind whose velocity is $v(x,t) = 2tx/(1+t^2) + 1 + t^2$ (from
Problem 1.4). Let the density of radioactive material be given by $\rho(x,t)$.

(a) Why should ρ evolve according to $\frac{\partial \rho}{\partial t} + v\frac{\partial \rho}{\partial x} = -\rho\frac{\partial v}{\partial x}$?

(b) Deduce the *characteristic* version of this equation:

$$\frac{d\rho}{dt} = -\rho\frac{\partial v}{\partial x} \quad \text{on curves } \mathcal{C} \text{ such that} \quad \frac{dx}{dt} = v.$$

Find these *curved* characteristics $\mathcal{C}$ and compare with $x^L(\xi, t)$.

(c) If the initial density of radioactive material is $\rho(x,0) = \rho_0(x)$, show
that at later times

$$\rho(x,t) = \frac{1}{1+t^2}\rho_0\left(\frac{x}{1+t^2} - t\right).$$

(d) How is this analysis changed if we take account of the fact that
radioactivity decays at a rate p (i.e. if any particular lump of material has a density of radioactive material ρ then $dr/dt = -p\rho$)?

DIFFICULT   **Problem 2.7** Investigate the effect on traffic flow of the *reaction time*
of car drivers.

(a) Hypothesise that the velocity at which cars travel is appropriate to
the car density a little earlier in time, i.e. $v = V[\rho(x, t-\tau)]$ where
τ is a "small" time. Argue that this car velocity-density relation
is approximately the same as

$$v = V(\rho) - \tau\frac{\partial \rho}{\partial t}V'(\rho).$$

(b) Substitute this into the continuity equation and derive that density
fluctuations must then (approximately) evolve according to

$$\frac{\partial \rho}{\partial t} + c(\rho)\frac{\partial \rho}{\partial x} = \tau\frac{\partial}{\partial x}\left(\frac{\partial \rho}{\partial t}\rho V'(\rho)\right).$$

(c) *Linearise* this equation by substituting $\rho = \rho_* + \hat{\rho}(x,t)$ and neglecting products of "small" quantities, $\hat{\rho}$, to derive that

$$\frac{\partial \hat{\rho}}{\partial x} + c_*\frac{\partial \hat{\rho}}{\partial x} = \tau\rho_* V_*'\frac{\partial^2 \hat{\rho}}{\partial x \partial t},$$

where $c_* = c(\rho_*)$ and $V'_* = V'(\rho_*)$. Look for solutions of the form $\hat{\rho} = \Re\{\exp(ikx + st)\}$, and show that the growth-rate s, dependent upon the wavelength of the perturbation $2\pi/k$, is $s = -ikc_*/(1 - ik\rho_* V'_* \tau)$.

(d) Hence show that, in heavy traffic ($\rho_* > \rho_m$), small sinusoidal perturbations of the uniform density traffic flow grow exponentially, according to

$$\hat{\rho} = \cos\left[k(x - c_* t)\right] \exp\left(k^2 \rho_* c_* V'_* \tau t\right)$$

Hint: once again use the fact that τ is "small", and consider the sign of $c_* V'_*$. How would this appear in the traffic flow?

2.3   Aggregation of slime mold amoebae

If food is plentiful, slime mold amoebae continually feed and multiply by mitosis, and are more-or-less evenly distributed in space. When the food supply becomes exhausted, the disappearance triggers certain changes in the amoebae. After a few hours the amoebae begin to aggregate into a number of collection points, typically a few hundred micrometres apart. The amoebae which have collected at a given point then form a multicelled slug, consisting of up to 10^5 *amoebae*, which moves as a unit. The slug eventually stops and releases spores from which amoebae appear when conditions are favourable, and the cycle begins again. What is responsible for the organised aggregation of the amoebae? This is an important question [9, p21], for "purposeful" movements occur frequently in developmental processes within organisms.

Also, what determines the time of onset of this organised aggregation? What determines the spacing of the aggregation centres? Can we quantify this process?

It has been discovered that the amoebae move preferentially toward relatively high concentrations of a chemical (AMP) which they themselves secrete. Presumably, aggregation is caused by the fact that the amoebae move up a gradient of attractant. We shall proceed to devise a simple mathematical model to describe the process; see [7, §1.3] for more details. If the analysis of this model is encouraging, then one could add more detail later, but we shall not.

2.3.1   The mathematical model

A one-dimensional model is sufficient to show the general results; a three-dimensional model would be better, but not significantly so in-so-far as we go. We model the amoebae and attractant as *two* intermingled interacting one-dimensional continua. Let $\rho(x,t)$ denote the density of amoebae and $a(x,t)$ the concentration of the attractant.

ASIDE

> The model we develop is about the simplest which is relevant to aggregation; however, it is more complicated than the car traffic model. It is curious that "simple" amoebae need a more complicated model than the "highly sophisticated" system of cars and drivers! Why?

Amoebae are conserved, as births and deaths are not significant during the relatively short time of aggregation, and so their density ρ must satisfy the continuity equation (2.1). The velocity with which they move is dependent upon the gradient of the attractant, and so we propose that $v = \chi \frac{\partial a}{\partial x}$, where χ is a constant which gives the strength of this *chemotaxis*. However, the amoebae also move around randomly as individuals. This random movement has the effect of making an additional contribution to the flux of amoebae past a given point (formerly just ρv) of $-\mu \frac{\partial \rho}{\partial x}$.

ASIDE

> This type of term in the flux is identical to that for almost all types of *diffusion*, see [7, §3.3]; individual microscopic random movements cause a macroscopic movement of the material. To see this here, suppose the elements or atoms of a continuum make small random movements. Imagine slicing the continuum into small slices of length δx, then in each slice i there will be a population of particles of (approximately) $N_i = \rho(x_i)\delta x$. Suppose that during a time interval δt a proportion p of the particles in slice i hop into the next slice to the right, slice $i+1$, and the same proportion (since direction is random), hop into the next slice to the left, slice $i-1$, and a proportion $1-2p$ stay in slice i. Now, look at the net flux of particles from slice i to slice $i+1$: in the time interval δt; the *net* number of particles moving from slice i to slice $i+1$ is just $n = pN_i - pN_{i+1}$. Thus the rate at which particles flow from slice i to $i+1$ is the flux
>
> $$\begin{aligned} q = \frac{n}{\delta t} &= \frac{p}{\delta t}\left(N_i - N_{i+1}\right) \\ &= \frac{p\delta x}{\delta t}\left(\rho(x_i) - \rho(x_{i+1})\right) \\ &= \frac{p\delta x}{\delta t}\left(\rho(x_i) - \rho(x_i) - \delta x \left.\frac{\partial \rho}{\partial x}\right|_{x_i} - \cdots \right) \\ &\approx -\left(\frac{p\,\delta x^2}{\delta t}\right)\frac{\partial \rho}{\partial x}\,. \end{aligned}$$

Thus the macroscopic flux past any given point is just $q = -\mu \frac{\partial \rho}{\partial x}$ (as used), where the constant $\mu = \left(p\delta x^2/\delta t\right)$ depends upon the details of the microscopic processes leading to the diffusion and is best found by experiment.

Hence, from the continuity equation the change in amoebae density is governed by

$$\frac{\partial \rho}{\partial t} + \frac{\partial q}{\partial x} = \frac{\partial \rho}{\partial t} + \frac{\partial}{\partial x}\left(\chi \rho \frac{\partial a}{\partial x} - \mu \frac{\partial \rho}{\partial x}\right) = 0 \, ,$$

which can be rewritten as

$$\frac{\partial \rho}{\partial t} = \mu \frac{\partial^2 \rho}{\partial x^2} - \chi \frac{\partial}{\partial x}\left(\rho \frac{\partial a}{\partial x}\right) \, . \qquad (2.8)$$

This is simply a diffusion equation for the amoebae, $\frac{\partial \rho}{\partial t} = \mu \frac{\partial^2 \rho}{\partial x^2}$, but modified by the chemotaxis interaction $\chi \frac{\partial}{\partial x}\left(\rho \frac{\partial a}{\partial x}\right)$.

Now the attractant is not conserved as it is generated by the amoebae and is also destroyed by the action of an enzyme. However, the derivation of the continuity equation may be adapted to cater for this (see Problem 2.1), once the generation and destruction processes are quantified. Suppose that at any point the attractant is generated at a rate which is proportional to the number of amoebae near that point, that is $f\rho$ where f is some constant. And suppose that the attractant is destroyed (as in radioactivity or some other spontaneous process) at a rate proportional to the amount of attractant present, that is pa where p is some constant. Lastly, suppose that the attractant is not transported and so the only contribution to the flux of attractant past a given point is due to random molecular movements (diffusion), that is $-D\frac{\partial a}{\partial x}$ where D is a constant. Thus the change in attractant density is just governed by the continuity equation (2.2) for AMP

$$\frac{\partial a}{\partial t} + \frac{\partial}{\partial x}\left(-D\frac{\partial a}{\partial x}\right) = f\rho - pa \, ,$$

which may be rewritten as

$$\frac{\partial a}{\partial t} = f\rho - pa + D\frac{\partial^2 a}{\partial x^2} \, . \qquad (2.9)$$

This is also a diffusion equation, but modified by the generation and destruction terms.

2.3.2   The uniform steady state

It is very easy to find an exact solution of these two governing equations (2.8) and (2.9). It is the uniform solution

$$a = a_* \quad \text{and} \quad \rho = \rho_* \quad \text{where} \quad f\rho_* = pa_* \ .$$

This last condition is physically reasonable. It says that in the uniform state the secretion rate of the attractant must exactly balance the destruction rate. We identify this uniform state with the period prior to aggregation when the amoebae are feeding.

2.3.3   Onset of aggregation

We model the onset of aggregation as the breakdown of this uniform state due to the growth of small disturbances. Small disturbances are inevitably present. This is a classical example of *instability*. The idea behind the analysis of instability is this. Suppose that at some time, the state of the system is slightly disturbed from the uniform state. Will these small disturbances tend to disappear with the passage of time, or will they become more intense? In the former case we say that the uniform state is stable; in the latter, unstable. Unstable states will not be observed as disturbances are inevitable.

ASIDE   The following analysis is directly analogous to the analysis of the stability of fixed points for a system of first-order ordinary differential equations. The only difference is that here the analysis takes place in an infinite-dimensional vector space, the space of differentiable functions, instead of a finite-dimensional space. For example, here the "fixed point" of the dynamical equations (2.8–2.9) is just the uniform solution discussed in the previous section.

To perform a stability analysis, investigate the behaviour of "small" disturbances to the uniform state by substituting

$$\rho = \rho_* + \hat{\rho}(x,t) \quad \text{and} \quad a = a_* + \hat{a}(x,t),$$

where $\hat{\rho}$ and $\hat{a}$ are "small" quantities. Then *linearise* the equations by neglecting all products of "small" terms. The amoebae equation (2.8) becomes

$$
\begin{aligned}
\frac{\partial \hat{\rho}}{\partial t} &= \mu \frac{\partial^2 \hat{\rho}}{\partial x^2} - \chi \frac{\partial}{\partial x}\left((\rho_* + \hat{\rho})\frac{\partial \hat{a}}{\partial x} \right) \\
&= \mu \frac{\partial^2 \hat{\rho}}{\partial x^2} - \chi \frac{\partial}{\partial x}\left(\rho_* \frac{\partial \hat{a}}{\partial x} \right) - \chi \frac{\partial}{\partial x}\left(\hat{\rho}\frac{\partial \hat{a}}{\partial x} \right),
\end{aligned}
$$

where the last term on the right-hand-side may be neglected as it is a product of small terms. The attractant equation (2.9) becomes

$$
\begin{aligned}
\frac{\partial \hat{a}}{\partial t} &= f(\rho_* + \hat{\rho}) - p(a_* + \hat{a}) + D\frac{\partial^2 \hat{a}}{\partial x^2} \\
&= (f\rho_* - pa_*) + f\hat{\rho} - p\hat{a} + D\frac{\partial^2 \hat{a}}{\partial x^2} \, ,
\end{aligned}
$$

where the first term on the right-hand-side is zero by the balance of the uniform state. These equations then reduce to

$$
\frac{\partial \hat{\rho}}{\partial t} = \mu \frac{\partial^2 \hat{\rho}}{\partial x^2} - \chi\rho_* \frac{\partial^2 \hat{a}}{\partial x^2} \quad \text{and} \quad \frac{\partial \hat{a}}{\partial t} = f\hat{\rho} - p\hat{a} + D\frac{\partial^2 \hat{a}}{\partial x^2} \, . \tag{2.10}
$$

This is a pair of linear, partial differential equations with constant co-efficients for the densities $\hat{\rho}$ and $\hat{a}$.

We guess that there are solutions of the form

$$
\hat{\rho} = \beta \cos(kx)e^{st} \quad \text{and} \quad \hat{a} = B \cos(kx)e^{st}, \tag{2.11}
$$

A standard guess for DEs with constant coefficients

where β and B are constant "amplitudes", and s is the growth-rate of such a disturbance of wavelength $2\pi/k$. From the differential equations we find equations which k, s, β and B must satisfy.

ASIDE

To explain the relevance of the above guess, imagine the amoebae are in a box of some length L. An initial perturbation $\hat{\rho}(x,0)$ and $\hat{a}(x,0)$ may written as Fourier series

$$
\hat{\rho}(x,0) = \sum_{n=0}^{\infty} \beta_n \cos\left(\frac{n\pi}{L}x\right) \quad \text{and} \quad \hat{a}(x,0) = \sum_{n=0}^{\infty} B_n \cos\left(\frac{n\pi}{L}x\right)
$$

for some coefficients β_n and B_n. Thus we are interested in what happens to perturbations with spatial dependence $\cos(kx)$ where k could take on any value, as $k = n\pi/L$ for arbitrary integer n and we do not really know what L is anyway. Given solutions as posed in (2.11) then the amoebae and AMP will evolve according to

$$
\begin{aligned}
\hat{\rho}(x,t) &= \sum_{n=0}^{\infty} \beta_n \cos\left(\frac{n\pi}{L}x\right) \exp(s_n t) \\
\text{and} \quad \hat{a}(x,t) &= \sum_{n=0}^{\infty} B_n \cos\left(\frac{n\pi}{L}x\right) \exp(s_n t)
\end{aligned}
$$

where s_n is the appropriate growth-rate for a wavenumber of $k = n\pi/L$. Thus: if *all* $s_n < 0$ then all factors $\exp(s_n t)$ decrease to zero, so must all the perturbations, and hence the uniform state is stable; contrariwise if $\exists s_n > 0$ then at least one factor $\exp(s_n t)$ increases with t, so must the corresponding perturbation, and the uniform state is unstable.

Now substitute the guess (2.11) into the equations (2.10), and find

$$s\beta \cos(kx)e^{st} = -\mu k^2 \beta \cos(kx)e^{st} + \chi\rho_* k^2 B \cos(kx)e^{st}$$
$$sB \cos(kx)e^{st} = f\beta \cos(kx)e^{st} - pB \cos(kx)e^{st} -$$
$$-Dk^2 B \cos(kx)e^{st} .$$

Divide by $\cos(kx)\exp(st)$ and rearrange to

$$\left[\begin{array}{cc} -\mu k^2 & \chi\rho_* k^2 \\ f & -p - Dk^2 \end{array} \right] \left[\begin{array}{c} \beta \\ B \end{array} \right] = s \left[\begin{array}{c} \beta \\ B \end{array} \right]$$

The interesting thing here is that for any given wavenumber k this is just an eigen-problem for eigenvalues s and corresponding eigenvectors $(\beta, B)^T$. It is straightforward eigen-analysis to derive that *nontrivial* solutions exist only if the growth-rate (eigenvalue) s is given by the quadratic (characteristic) equation

$$s^2 + bs + c = 0 , \qquad (2.12)$$

where $b = p + (\mu + D)k^2 > 0$ and $c = \mu k^2(p + Dk^2) - \chi\rho_* fk^2$. Thus there are two families of solution with growth-rates $s = s^{\pm}(k) = \left(-b \pm \sqrt{b^2 - 4c}\right)/2$. However, we are not interested in the details of these growth-rates, all we need to know is when one of them becomes positive as only then does the uniform state become unstable.

We are interested in the zeros of the quadratic $s^2 + bs + c$. Since the coefficient of the s^2 terms is positive, namely 1, the parabola it describes bends upwards as shown in Figure 2.15. Since $b > 0$ the slope of the parabola, $b + 2s$, is > 0 at $s = 0$ and so the curve must be increasing as it crosses the vertical axis. The only possibilities for this quadratic are as shown in Figure 2.15 and it is clear that there is a positive zero of the quadratic *if and only if* the coefficient $c < 0$; only then will there exist a growing mode (2.11).

Now the coefficient

$$c = (\mu p - \chi\rho_* f) k^2 + \mu Dk^4 ,$$

is a simple quartic in k, where the wavenumber k is arbitrary—a typical disturbance to the uniform state will involve a wide range of k. Note that the coefficient of k^4 is always positive and this term dominates the large k dependence, while the coefficient of k^2 may be positive or negative

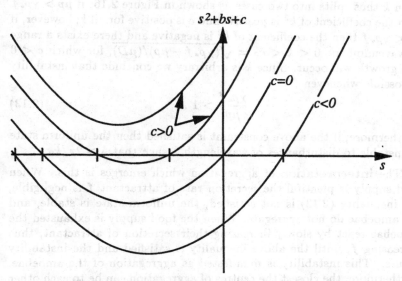

Figure 2.15: The only possibilities for the family of quadratics (2.12) showing when a positive zero may occur.

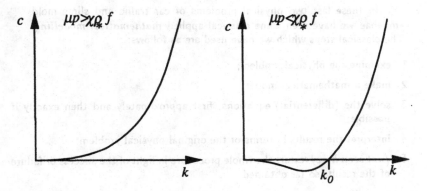

Figure 2.16: The wavenumber dependence of the coefficient c showing that $c < 0$ only when the inequality (2.13) is satisfied.

and this term dominates the small k dependence. The dependence of c upon k thus splits into two cases as shown in Figure 2.16: if $\mu p > \chi \rho_* f$ then the coefficient of k^2 is positive and c is positive for all k; however, if $\mu p < \chi \rho_* f$ then the coefficient of k^2 is negative and there exists a range of wavenumbers, $0 < k < k_0 = \sqrt{(\chi \rho_* f - \mu p)/(\mu D)}$, for which $c < 0$ and growth will occur. Since k is arbitrary we conclude that instability is possible whenever

$$\frac{\chi \rho_* f}{\mu p} > 1 \ . \tag{2.13}$$

Furthermore, if the above constraint is satisfied then the uniform state is unstable to disturbances of wavelength λ such that $\lambda > 2\pi/k_0$.

The **interpretation** of aggregation which emerges is this. When food supply is plentiful the secretion rate of attractant f is negligible, the inequality (2.13) is not satisfied, the uniform state is stable, and the amoebae do not aggregate. When the food supply is exhausted the amoebae react by slowly increasing their secretion of attractant, thus increasing f, until the above inequality is satisfied and the instability occurs. This instability is manifested as aggregation of the amoebae. Furthermore, the closest the centres of aggregation can be to each other is the shortest wavelength of the growing modes. Thus the aggregation centres must be spaced further apart than $2\pi/k_0$.

Later stages of the aggregation are not described by this linearisation, and a more sophisticated analysis is needed. However, this stability analysis is always the starting point.

ASIDE

In these last two physical problems of car traffic and slime mold amoebae we have seen some classical applied *mathematical modelling*. The classical steps which we have used are as follows:

1. examine the physical problem;

2. make a mathematical model;

3. solve the (differential) equations, first approximately and then exactly if possible;

4. interpret the results in terms of the original physical problem;

5. *(not seen here)* restart the whole procedure in light of the success or failure of the results so far obtained.

Exercises

EASY   **Problem 2.8** For the following two distributions of density of some material:

$$(i) \quad \rho = t^{-1/2} \exp\left(-x^2/4\mu t\right) \; ; \qquad (ii) \quad \rho = -2x \; .$$

(a) Sketch ρ as a function of x for a couple of different times.

(b) Verify that ρ satisfies the diffusion equation: $\frac{\partial \rho}{\partial t} = \mu \frac{\partial^2 \rho}{\partial x^2}$.

(c) What is the flux of material, q, past any point x at any time?

Problem 2.9 A pendulum on a rigid rod oscillates according to the EASY equation

$$\frac{d^2\theta}{dt^2} + \sin\theta = 0,$$

where θ is the *angle* of the pendulum from the vertical.

(a) Show that $\theta = 0$ (the pendulum hanging straight down) and $\theta = \pi$ (the pendulum balanced vertically up) are possible equilibrium (or steady) solutions of the pendulum equation.

(b) By *linearisation* show that the solution $\theta = \pi$ is unstable, while the solution $\theta = 0$ is stable; at least, show that it is not unstable.

> Thus the solution $\theta = \pi$, the pendulum balancing vertically upwards, ASIDE
> is theoretically possible, but it is never seen in practise as it is unstable. It
> is interesting to note that if we apply a vibration to the pivot of a pendulum
> then this fixed point, $\theta = \pi$, may be made stable! see Problem 11.21 in
> [1, p568].

Problem 2.10 Consider our model of slime mold amoebae when the DIFFICULT parameters have the value $d = D = \mu = \chi = 1$ and the initial density of amoebae is $\rho_* = 1$.

(a) Show that the uniform equilibrium solution is $\rho = 1$ and $a = f$, where f is the secretion rate of AMP.

(b) For the three values of the secretion rate $f = 0$, $f = 1$ and $f = 3$; sketch the graphs of the two growth rates, $s = s_1(k)$ and $s = s_2(k)$ of small perturbations (of wavelength $2\pi/k$) to the uniform state (as a function of k for $0 \le k \le 2$). For which values of f does there exist a range of k with unstable solutions ($s_i > 0$)?

(c) Suppose the amoebae are placed in a box of length π so that the linearised solutions can be expressed as

$$\hat{\rho} = \sum_{n=0}^{\infty} \left[\alpha_n e^{s_1(n)t} + \beta_n e^{s_2(n)t} \right] \cos(nx),$$

$$\hat{a} = \sum_{n=0}^{\infty} \left[A_n e^{s_1(n)t} + B_n e^{s_2(n)t} \right] \cos(nx).$$

From the governing linearised equations, what relation must exist between the coefficients α_n and A_n, and between β_n and B_n?

MEDIUM

Problem 2.11 A country is to establish a zone in the sea adjacent to its coast in which no fishing is to be allowed. This is to be done to protect a species of fish which, in the absence of fishing, would grow exponentially due to births which are proportional to the density of the fish. Outside the zone, which extends from the shore to a distance L out to sea, *all* the fish are harvested by deep-sea trawlers. The fish move about at random in the direction from the shore to the zone's boundary and so effectively diffuse in this direction. If L is too small, the effectively immediate destruction of the fish which wander out of the protection zone will cause a loss of fish that may exceed their ability to reproduce; this would lead to eventual extinction. Formulate a model of the dynamics based on the conservation of fish, their breeding and their movement. Find the minimum zone width L that prevents the extinction of the fish by the trawlers. Note that a boundary condition is obtained from realising that the fish cannot swim on-shore (i.e. there is no flux of fish past the shoreline). Another boundary condition is obtained by realising that there are effectively no fish outside the protection zone, and this applies all the way to the edge of the zone.

Chapter 3

Balance of momentum

For continua composed of "dumb" molecules which obey Newton's second law, the fact that each individual molecule obeys $F = ma$ is reflected in the macroscopic behaviour of the material. Through this observation we derive a further dynamical equation, in addition to the continuity equation (2.1), which governs the movement of most classical continuous materials.

3.1 The momentum equation

A more fundamental form of Newton's second law is simply $\frac{dp}{dt} = F$; that is, the time rate of change of momentum is equal to the applied force. Because this form of the law applies even when the mass of the molecule is changing, it is more suitable for our use in a continuum model; this is because the continual movement of the material constantly alters which molecules are under consideration. Thus we first investigate the momentum of a continuum and the nature of forces on a continuum.

3.1.1 Momentum and momentum flux

Consider a one-dimensional continuum with density field ρ and velocity field v. The momentum density of such a continuum at any point is just the average of the molecular momentum around that point, as explained for the mass density in Section 1.1. However, under the hypothesis that the randomness in the location and masses of the molecules are not

47

correlated with the randomness in the velocity of the molecules, the
momentum density is simply

$$p = \rho v \ .$$

ASIDE   Actually this equation is best taken as a definition of the average-
velocity field—the density and the momentum density fields being defined
by the averaging process described in Section 1.1, and the average velocity
then being $v = p/\rho$.

We also have to quantify how this momentum moves in the continuum. It is carried by the molecules of the continuum and so the local
momentum density moves with the local continuum velocity. Hence, the
rate at which momentum is being carried to the right past any point,
the **momentum flux**, is given by

$$pv = (\rho v)v = \rho v^2 \ .$$

Note that if the velocity v is negative, then this momentum flux is still
positive because then negative momentum is being carried to the left,
which ends up as a positive momentum flux.

3.1.2   Forces

The other quantity involved in Newton's second law is the force. In a
continuum there are two distinct types of forces—and although you are
all familiar with their effects you may not realise their presence.

The first type of force are those *external forces* which act on each
and every molecule of the continuum, gravity for example. These **body
forces**, as they are called, clearly act throughout the continuum and
are usually of a known nature. The body force per unit length will be
denoted by F. For example: gravity pulling to the left (towards negative
x) is a body force $F = -g\rho$; fluid flowing through a pipe encounters a
frictional resistance due to friction with the walls of the pipe which
may be modelled by a body force $F = -Cv$ where C is some constant
coefficient of drag.

Gravity will
be our main
body force

ASIDE   It is gravity which holds the earth's atmosphere down near the surface
of the earth, as examined in Problem 3.4.

The second type of force is that which is transmitted through the
continuum by a molecule acting upon its neighbouring molecules, and
being acted upon in return, by Newton's third law. It is these forces
which make the whole of any solid body move when we hold and move

just one small part of it. Such a force will be called a **stress** and will be denoted by σ. At an arbitrary point X of the continuum the material just to the right of X *either* pulls the material on the left of X (and gets pulled in return)—this **tension** corresponds to positive stress, *or* pushes the material on the left (and gets pushed in return)—this **compression** corresponds to negative stress.

3.1.3 Conservation of momentum

Newton's second law states that the only way to change momentum is by applying forces, so it is really a statement that momentum is conserved. Just as the conservation of mass lead to the continuity equation in Section 2.1.2, we examine a slice of the continuum to derive an equation which reflects the conservation of momentum and its generation by forces.

Consider any fixed slice of a continuum, say the interval $[a, b]$ as shown in Figure 3.1. The total momentum in the slice is simply the in-

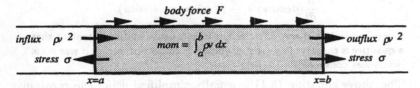

Figure 3.1: A slice of a continuum to investigate the conservation of momentum

tegral of the momentum density, namely $\int_a^b \rho v\,dx$. Momentum can flow into or out of the slice across the ends at the rate ρv^2, the momentum flux. Finally, momentum can be generated (or destroyed) in the slice by the applied forces: a body force F throughout the slice; and stresses σ applied across the ends of the slice. Starting from a modified statement of Newton's second law we deduce

(rate of momentum increase in slice)

= (rate of momentum influx across ends) + (net applied forces)

$$\Rightarrow \quad \frac{\partial}{\partial t} \int_a^b \rho v\,dx$$

$$= \left(\rho(a,t)v(a,t)^2 - \rho(b,t)v(b,t)^2\right) + \left(\int_a^b F\,dx + \sigma(b,t) - \sigma(a,t)\right)$$

The physical processes in this slice.

$$\Rightarrow \quad \int_a^b \frac{\partial}{\partial t}(\rho v)\, dx = -\left[\rho v^2\right]_{x=a}^b + \int_a^b F\, dx + \left[\sigma\right]_{x=a}^b$$

$$\Rightarrow \quad \int_a^b \frac{\partial}{\partial t}(\rho v)\, dx = -\int_a^b \frac{\partial}{\partial x}\left(\rho v^2\right) dx + \int_a^b F\, dx + \int_a^b \frac{\partial \sigma}{\partial x}\, dx$$

$$\Rightarrow \quad \int_a^b \left\{ \frac{\partial}{\partial t}(\rho v) + \frac{\partial}{\partial x}\left(\rho v^2\right) - \left(F + \frac{\partial \sigma}{\partial x}\right) \right\} dx = 0$$

Since this identity is true for all slices in the continuum, for all $a < b$, then by the slicing theorem in Section 2.1.1 the integrand must be zero for all x. Thus the following equation

$$\frac{\partial}{\partial t}(\rho v) + \frac{\partial}{\partial x}\left(\rho v^2\right) = F + \frac{\partial \sigma}{\partial x} \ , \tag{3.1}$$

reflecting conservation of momentum, must hold throughout the continuum.

ASIDE

Observe that this equation has exactly the same form as the generalised continuity equation (2.2). This form of governing equation

$$\frac{\partial}{\partial t}(\text{density}) + \frac{\partial}{\partial x}(\text{flux}) = (\text{generation})$$

is very general; it always arises as the consequence of any statement that a quantity is conserved—here momentum is conserved, earlier it was mass which was conserved.

The above equation (3.1) is usually simplified using the continuity equation. This is done by expanding the derivatives on the left-hand-side and rearranging; equation (3.1)

$$\Rightarrow \quad \frac{\partial \rho}{\partial t}v + \rho\frac{\partial v}{\partial t} + \frac{\partial(\rho v)}{\partial x}v + \rho v\frac{\partial v}{\partial x} = F + \frac{\partial \sigma}{\partial x}$$

$$\Rightarrow \quad v\left(\frac{\partial \rho}{\partial t} + \frac{\partial(\rho v)}{\partial x}\right) + \rho\left(\frac{\partial v}{\partial t} + v\frac{\partial v}{\partial x}\right) = F + \frac{\partial \sigma}{\partial x} \ .$$

The first bracketed term on the left-hand-side is identically zero by the continuity equation (2.1) and hence the following **momentum equation**

$$\rho\left(\frac{\partial v}{\partial t} + v\frac{\partial v}{\partial x}\right) = F + \frac{\partial \sigma}{\partial x} \ . \tag{3.2}$$

must hold at all points in the continuum.

This momentum equation has an interesting and curiously simple form. The bracketed term on the left-hand-side of equation (3.2) is recognised as the material derivative of the velocity and so it may be

An easily remembered form.

written as

$$\rho \frac{Dv}{Dt} = F + \frac{\partial \sigma}{\partial x}.$$

In this form the equation asserts that (mass density) × (particle acceleration) = (total force), which is directly analogous to Newton's second law $ma = F$! It is curious that after so much work we end up with an equation which is so very close to what we started with.

Exercises

Problem 3.1 This problem illustrates the nature of stress.   EASY

(a) Consider a tower of 4 blocks, each of mass m, in a gravitational field in *equilibrium*, that is there is no change in time, as shown in Figure 3.2. Write down the force which block $i + 1$ exerts on

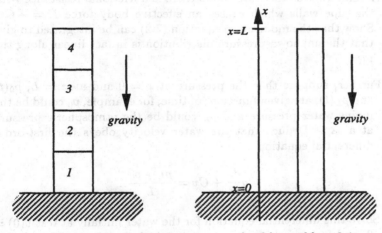

Figure 3.2: the two towers examined in problem 3.1.

block i for $i = 1, 2, 3$ (assume that no force is exerted on the top of block 4 and take an upward force to be positive). Also find the force which block 1 exerts on the ground.

(b) Consider a rod of length L and constant density ρ standing on end in *equilibrium*, that is $v = 0$ and there is no change in time, in a gravitational field as shown in Figure 3.2. Use the momentum equation to find the distribution of stress σ in the rod; assume that no force (stress) is exerted on the top of the rod. Compare the stress distribution to the answer for part (a).

MEDIUM   **Problem 3.2** Derive an appropriately modified momentum equation
for the system of Problem 2.1, where material is being continually added
to the continuum at a rate $r(x,t)$ and which moves with mass flux
$q = \rho v$. Assume a velocity of $u(x,t)$ for the material which is being
added at the point (x,t).

MEDIUM   **Problem 3.3 Simple fluid flow in a pipe.** As a first approximation
in fluids we may take the stress $\sigma = -p$ where p is the fluid's pressure.
Consider a pipe of length L, stretching from $x = 0$ to $x = L$, filled
with water which has a constant density (to a good approximation),
$\rho = $ const.

(a) Use the continuity equation (2.1) to show that the velocity of the
     water is a function of time only, *i.e.* $v = v(t)$.

(b) Assume that as the water flows there is a frictional resistance with
     the pipe walls which causes an effective body force $F = -Cv$.
     Show that the momentum equation (3.2) can be integrated to give
     that the unknown pressure distribution is in fact linear along the
     pipe.

(c) Further, suppose that the pressure at $x = 0$ and at $x = L$, $p_0(t)$
     and $p_L(t)$, are given functions of time; for example, p_0 could be the
     mains water pressure and p_L could be the atmospheric pressure
     at a tap. Deduce that the water velocity obeys the first-order
     differential equation

$$\rho \frac{dv}{dt} + Cv = \frac{p_L - p_0}{L} \ .$$

(d) Solve this differential equation for the water initially at rest $v(0) =$
     0, and for a pressure drop $p_L - p_0$ which is constant (to simulate
     the opening of a tap at time $t = 0$).

3.2   Ideal gas dynamics

Some of the simpler materials in our world are gases, for example the
air we breathe, to which we now turn to provide a familiar example
of the use of the combination of both the continuity equation and the
momentum equation. Here we concentrate on the propagation of sound.
The behaviour of solids and liquids is discussed in the next two chapters.

3.2.1 The equation of state

So far we have two equations, the continuity equation (2.1) and the momentum equation (3.2), for three unknowns, ρ, v and σ. Note that the body force F is assumed to have a known nature depending upon what is causing it. Since there is one more unknown than equations, we need another equation to "close the problem", just as for car traffic. Here this comes from school science in the form of the equation of state for ideal gases:

$$pV = nRT \; ,$$

where T is the gas temperature, R is a constant, n is the number of *moles* of gas, V is the volume containing the gas, and p is the gas pressure.

For our purposes, write this equation as

$$p = \rho RT \; ,$$

where $\rho = n/V$ is clearly the density of the gas. Furthermore, assume that this equation holds not only for a large container of gas, but also for all points in the gas. That is, physically we assume it holds in all small regions which are as big as an averaging length; mathematically it then holds at all "points" in the gas.

But what is the pressure p? Pressure is a force exerted by the gas across or on a surface. Thus pressure is very much like a stress, in fact $p = -\sigma$ and so the equation of state becomes

$$\sigma = -\rho RT \; .$$

ASIDE

Actually, for a real gas the stress is more like $\sigma = -p +$ (friction terms), which accounts for some irreversible processes in the gas. However, here we shall just consider the reversible dynamics of an ideal gas.

What about the temperature T? Recall that compressing a gas makes it hotter; for example, the end of a bicycle pump becomes hotter when you pump up a tyre, especially a high pressure tyre. Conversely, expanding a gas generally cools it; for example, the air coming out of a car tyre if you let it down is cool. In fact, to a good approximation the temperature is proportional to a power of the pressure, namely $T = K\rho^{\gamma-1}$ where K and γ are constants which depend upon the gas under consideration. Thus

$\gamma \approx 7/5$ for air

$$\sigma = -\rho RK\rho^{\gamma-1} = -RK\rho^{\gamma} \; ,$$

which is more usually written in the form

$$\sigma = -\left(\frac{k^2}{\gamma}\right)\rho^\gamma \,, \tag{3.3}$$

where γ and $k = \sqrt{\gamma R K}$ are constants of the gas, which may be determined experimentally.

This last equation supplies the third equation we need for the three unknowns: ρ, v and σ. However, it is convenient to eliminate the stress σ from the equations to give the two equations

$$\frac{\partial \rho}{\partial t} + \frac{\partial}{\partial x}(\rho v) = 0 \tag{3.4}$$

$$\rho\left(\frac{\partial v}{\partial t} + v\frac{\partial v}{\partial x}\right) = F - k^2 \rho^{\gamma-1}\frac{\partial \rho}{\partial x} \,, \tag{3.5}$$

for the two unknowns ρ and v.

3.2.2   The wave equation

For the next two sections we take the body forces to be zero, $F = 0$, that is we assume they have a negligible influence compared to the discussed phenomena. The most basic process in an ideal gas is the propagation of sound, density-velocity waves, and it is this which we now describe.

A fixed point of the dynamical equations   We first identify a state of rest, or equilibrium,

$$v = 0 \quad \text{and} \quad \rho = \rho_* \,,$$

where ρ_* is any constant, say the density of the atmosphere near the earth's surface.

Then the behaviour of disturbances to this basic state are found by linearisation. As in previous chapters assume

$$v = \hat{v}(x,t) \quad \text{and} \quad \rho = \rho_* + \hat{\rho}(x,t) \,,$$

where the perturbations $\hat{v}$ and $\hat{\rho}$ are "small" in the sense that we neglect products of "small" terms. Substituting into the continuity equation (3.4) deduce

$$\frac{\partial \hat{\rho}}{\partial t} + \frac{\partial}{\partial x}\left(\rho_* \hat{v} + \hat{\rho}\hat{v}\right) = 0 \quad \Rightarrow \quad \frac{\partial \hat{\rho}}{\partial t} + \rho_*\frac{\partial \hat{v}}{\partial x} \approx 0 \,.$$

Substituting into the momentum equation (2.5) gives

$$(\rho_* + \hat{\rho})\left(\frac{\partial \hat{v}}{\partial t} + \hat{v}\frac{\partial \hat{v}}{\partial x}\right) = -k^2(\rho_* + \hat{\rho})^{\gamma-1}\frac{\partial \hat{\rho}}{\partial x}$$

$$\Rightarrow \quad \rho_*\frac{\partial \hat{v}}{\partial t} + \rho_*\hat{v}\frac{\partial \hat{v}}{\partial x} + \hat{\rho}\frac{\partial \hat{v}}{\partial t} + \hat{\rho}\hat{v}\frac{\partial \hat{v}}{\partial x} = -k^2\rho_*^{\gamma-1}\left(1 + \frac{\hat{\rho}}{\rho_*}\right)^{\gamma-1}\frac{\partial \hat{\rho}}{\partial x}$$

$$\Rightarrow \quad \rho_*\frac{\partial \hat{v}}{\partial t} \approx -k^2\rho_*^{\gamma-1}\frac{\partial \hat{\rho}}{\partial x}$$

in which we have used the Taylor's series approximation for the binomial

$$\left(1 + \frac{\hat{\rho}}{\rho_*}\right)^{\gamma-1} = 1 + (\gamma-1)\frac{\hat{\rho}}{\rho_*} + \frac{(\gamma-1)(\gamma-2)}{2}\frac{\hat{\rho}^2}{\rho_*^2} + \cdots \approx 1 \ .$$

These two linearised equations describe the propagation of waves. To see this, differentiate the linearised continuity equation with respect to t to find

$$\frac{\partial^2 \hat{\rho}}{\partial t^2} + \rho_*\frac{\partial^2 \hat{v}}{\partial x \partial t} = 0$$

which, upon substituting for $\frac{\partial \hat{v}}{\partial t}$ from the linearised momentum equation, becomes

$$\frac{\partial^2 \hat{\rho}}{\partial t^2} + \frac{\partial}{\partial x}\left(-k^2\rho_*^{\gamma-1}\frac{\partial \hat{\rho}}{\partial x}\right) = 0 \ .$$

Rearranging, this equation is

$$\frac{\partial^2 \hat{\rho}}{\partial t^2} = \underbrace{k^2\rho_*^{\gamma-1}}_{c_*^2}\frac{\partial^2 \hat{\rho}}{\partial x^2} \ , \tag{3.6}$$

which is the **wave equation** describing waves of density fluctuations (i.e. sound) travelling to the left or to the right with wave speed $c_* = k\rho_*^{(\gamma-1)/2}$. This is most easily seen by D'Alembert's solution: $\hat{\rho} = f(x - c_*t) + g(x - c_*t)$ where f and g are any twice differentiable functions which in any specific situation are given by the initial and boundary conditions.

Any sound wave also involves velocity fluctuations that travel with the wave. For example, from the linearised continuity equation and D'Alembert's solution we find the velocity field is given by $\hat{v} = \frac{c_*}{\rho_*}[f(x - c_*t) - g(x - c_*t)]$ which describes velocity fluctuations travelling in phase with the density fluctuations.

3.2.3   Loud unidirectional sound

We now "specialise" the governing equations of gas dynamics so that
Exact solution they describe sound which propagates in only one direction. The governing equations then simplify and are solved by the method of characteristics developed for car traffic flow in Section 2.2.4.

Once again assume that body forces are negligible, that is $F = 0$. The simplification is to consider only those solutions of the governing equations (3.4–3.5) for which

$$\rho = R(v) \; ,$$

where $R(v)$ is some specific but as yet unknown function. We will find that this restriction is consistent with the original equations only for some particular choice for $R(v)$.

For simplicity take the gas constant $\gamma = 1$. For other values of γ the details to follow will be different, but the qualitative results are the same, see Problem 3.4.

Substituting the assumption $\rho = R(v)$ into the continuity equation (3.4) and using the chain rule

$$\Rightarrow \quad R'(v)\frac{\partial v}{\partial t} + R'(v)\frac{\partial v}{\partial x}v + R(v)\frac{\partial v}{\partial x} = 0$$

$$\Rightarrow \quad \left(\frac{\partial v}{\partial t} + v\frac{\partial v}{\partial x}\right)R' = -R\frac{\partial v}{\partial x}$$

$$\Rightarrow \quad \frac{\partial v}{\partial t} + v\frac{\partial v}{\partial x} = -\left(\frac{R}{R'}\right)\frac{\partial v}{\partial x} \; .$$

Similarly substituting into the momentum equation (3.5) leads to

$$\begin{aligned}
\frac{\partial v}{\partial t} + v\frac{\partial v}{\partial x} &= -k^2\frac{1}{\rho}\frac{\partial \rho}{\partial x} \\
&= -k^2\frac{1}{R}R'(v)\frac{\partial v}{\partial x} \\
&= -\left(k^2\frac{R'}{R}\right)\frac{\partial v}{\partial x} \; .
\end{aligned}$$

The assumption that $\rho = R(v)$ is only consistent with both the continuity and momentum equations if these two equations become identical; this clearly can only occur if the bracketed factors are the same, that is

$$\frac{R}{R'} = k^2\frac{R'}{R} \; .$$

This is simply a separable first-order differential equation for $R(v)$ which is solved via the following steps

$$\left(\frac{R'}{R}\right)^2 = \frac{1}{k^2}$$

$$\Rightarrow \quad \int \frac{dR}{R} = \int \frac{\pm 1}{k}\, dv$$

$$\Rightarrow \quad \log R = \pm v/k + A$$

$$\Rightarrow \quad R = \exp(\pm v/k + A)$$

where A is an arbitrary constant of integration.

Thus, taking

$$\rho = \rho_* \exp(v/k) \, , \qquad (3.7)$$

where $\rho_* = e^A$ is an arbitrary constant (ρ_* happens to be the density whenever the gas is stationary) then both the continuity equation and the momentum equation reduce to the same equation. From the continuity equation we find

The $\exp(-v/k)$ case is similar but with sign changes.

$$\frac{\partial v}{\partial t} + v\frac{\partial v}{\partial x} = -\frac{\rho_* \exp(v/k)}{\rho_* \frac{1}{k} \exp(v/k)}\frac{\partial v}{\partial x} = -k\frac{\partial v}{\partial x} \, ,$$

which is conveniently written as

$$\frac{\partial v}{\partial t} + (k + v)\frac{\partial v}{\partial x} = 0 \, . \qquad (3.8)$$

This is exactly the same equation as was solved for car traffic, namely equation (2.4), but here the coefficient function is just the linear function $c(v) = k+v$. This means that we can solve it in exactly the same manner, via the method of characteristics, to find a class of exact solutions of the ideal gas equations (3.4–3.5).

From the discussion of this sort of equation modelling car traffic we appreciate that this equation describes the propagation of density-velocity waves (sound) of arbitrary amplitude to the right; the case $\rho = \rho_* \exp(-v/k)$ describes left travelling waves. Observe that if v is small, equation (3.8) is approximately $\frac{\partial v}{\partial t} + k\frac{\partial v}{\partial x} = 0$ which has the simple solution $v = f(x - kt)$ describing waves travelling to the right, unchanging in form, with velocity k. In most situations the nonlinear term in the equation $v\frac{\partial v}{\partial x}$ induces "shocks" to eventually form; these shocks are more commonly known as **sonic booms**.

The linearised wave.

However, we are not inundated by sonic booms from the sound all   ASIDE around us. The reason is that, for ordinary levels of sound, the time-scale for the formation of the shock is so long that weak dissipative processes, ignored in our model, have time to dampen out the sound waves. It is only extra-ordinarily loud sound, such as that generated by a supersonic jet or a gunshot, which will generate a shock.

Exercises

MEDIUM

Problem 3.4 Consider the equations of ideal gas dynamics, equations (3.4–3.5)

(a) Take $\gamma = 7/5$ (suitable for air) and find the distribution of density of an atmosphere in *equilibrium* when it is acted on by a uniform gravitational field $F = -g\rho$ (assume $\rho = \rho_0$ at $x = 0$). Sketch graphs of the density and the stress as functions of height x.

(b) Take body forces to be negligible, $F = 0$, and $\gamma = 7/5$ (i.e. air) and find the equation governing those solutions of the ideal gas equations which satisfy some functional relation $\rho = R(v)$. The final equation has the same form as equation (3.8), but the coefficients are different.

MEDIUM

Problem 3.5 Consider an ideal gas with $\gamma = 1$ which is restricted to right-travelling waves described by equations (3.7–3.8).

(a) Show that the characteristic solution of equation (3.8) is $v = $ const. on straight lines $\frac{dx}{dt} = k + v$.

(b) Consider a semi-infinite pipe $0 < x < \infty$ filled with an ideal gas initially at rest, $v = 0$ at $t = 0$, with constant density $\rho = \rho_0$ and with parameter $k = 1$ (in some system of units). An ideal fan which is located at $x = 0$ blows in gas with velocity $v = \frac{1}{2\pi} \sin(\pi t)$ for $0 < t < 2$ and then stops blowing $v = 0$ for $t > 2$; this generates a short wave-pulse which travels down the pipe. Draw the graph of the characteristics and hence plot the velocity v over $0 \leq x \leq 4.5$ at times $t = 2$ and $t = 4$. Two shocks eventually form (as is typical of "sonic booms"); about when does the first shock form? Note that the domain of this problem is the first quadrant with characteristics emanating from both the x-axis (for $x \geq 0$) and the t-axis (for $t \geq 0$).

3.3 Quasi-linear partial differential equations

Spurred by the generation of shocks in car traffic and the desire to form more general solutions describing sound waves, it is interesting to develop a little general theory. This section is not needed for the development of continuum mechanics in the book, and is significantly harder. However, it will complete a more detailed picture of one-dimensional dynamics.

3.3.1 Shocks and weak solutions

To start with, reconsider the general first-order differential equation for an unknown $\rho(x,t)$

$$\frac{\partial \rho}{\partial t} + c(\rho, x, t)\frac{\partial \rho}{\partial x} = r(\rho, x, t) \qquad (3.9)$$

as could arise in modelling car traffic, but with a source of cars, $r(\rho, x, t)$, acknowledged in the equation. Such a source (or a sink if r is negative) may arise from cars entering or leaving the road from side streets or parking stations.

As described in Subsection 2.2.4 on car traffic, given that there exists a solution field $\rho(x,t)$ we may imagine finding **characteristic curves** $\mathcal{C}$ defined by

$$\frac{dx}{dt} = c\left[\rho(x,t), x, t\right] \ .$$

Then on a specific curve $\mathcal{C}$, described by $x = x(t)$ say, ρ has values which are a function of time alone: $\rho = \rho(x(t), t)$. Differentiating with respect to time, considering only the time evolution of ρ along the characteristic, we find

$$
\begin{aligned}
\frac{d\rho}{dt} &= \frac{\partial \rho}{\partial t} + \frac{\partial \rho}{\partial x}\frac{dx}{dt} \quad \text{by the chain rule} \\
&= \frac{\partial \rho}{\partial t} + c(\rho, x, t)\frac{\partial \rho}{\partial x} \quad \text{by the definition of } \mathcal{C} \\
&= r(\rho, x, t) \quad \text{by (3.9)}
\end{aligned}
$$

Thus the characteristic solution in full is to solve the pair of coupled ordinary differential equations

$$\boxed{\begin{aligned} \frac{d\rho}{dt} &= r(\rho, x, t) \\ \frac{dx}{dt} &= c(\rho, x, t) \end{aligned}} \qquad (3.10)$$

to give a characteristic curve $x(t)$ and the solution ρ on it. One way to appreciate how this solution works is to imagine a simple numerical scheme based on these equations. As shown in Figure 3.3, starting at $t = t_0 = 0$ with the characteristic through $x = x_0$ where the initial value of the unknown field is $\rho = \rho_0$, we then build up an approximate solution through the finite difference equations

$$
\begin{aligned}
t_{n+1} &= t_n + \Delta t \\
\rho_{n+1} &= \rho_n + r(\rho_n, x_n, t_n)\Delta t \\
x_{n+1} &= x_n + c(\rho_n, x_n, t_n)\Delta t
\end{aligned}
$$

Doing this for an appropriate variety of different values of x_0, with the corresponding ρ_0, builds a set of characteristics which cover the entire xt-plane and which thus implicitly gives the solution field $\rho(x, t)$.

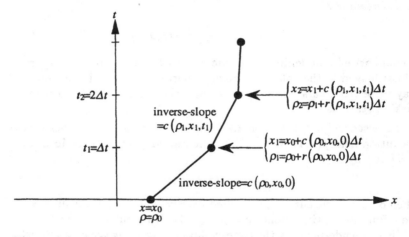

Figure 3.3: A diagram showing how a simple numerical scheme may be used to obtain approximate characteristic solutions to equation (3.9).

ASIDE   In the earlier car traffic examples $r = 0$ and c was independent of x and t, consequently all the characteristics were straight lines. But this is not the case in general.

However, as seen in the earlier analysis of car traffic, this characteristic solution leads to multi-valued solutions whenever the characteristic curves cross. For example, the evolution following a "bunching" together of a group of cars leads to the crossing characteristics of Figure 2.10 which predicts the multi-valued solution at time $t = t_3$ seen in Figure 2.11! But the process of averaging observations must give a single value for the density field ρ. This implies that there must be a transition (a shock) from the solution seen on the left to that on the right. But where is it located?

The answer to this question is found by observing that differentiability is lost at time $t = t_2$ when the x-slope of ρ first becomes vertical. Thus the differential form of car conservation is useless at, and after, this time. However, conservation is an extremely robust principle and the integral form of the conservation still applies. This enables us to discover the dynamics of a shock.

Reconsider the general equation (3.9) in the conservative form

$$\frac{\partial \rho}{\partial t} + \frac{\partial q}{\partial x} = r'(\rho, x, t) , \qquad (3.11)$$

where the flux $q = \int c(\rho, x, t) \, d\rho$ and $r' = r + q_x$ (in which the subscript x denotes that the differentiation is done considering ρ to be constant). This equation is equivalent to (3.9) as, according to (3.9),

$$\frac{\partial \rho}{\partial t} + \frac{\partial q}{\partial x} = \frac{\partial \rho}{\partial t} + q_\rho \frac{\partial \rho}{\partial x} + q_x = \frac{\partial \rho}{\partial t} + c \frac{\partial \rho}{\partial x} + q_x = r + q_x = r'$$

This differential equation arises from the integral conservation law

$$\frac{\partial}{\partial t} \int_a^b \rho \, dx + q(b, t) - q(a, t) = \int_a^b r' dx \qquad (3.12)$$

which is used when the differential form breaks down.

Suppose the differential equation (3.11) holds everywhere except for a discontinuity across a **shock** located at some time-dependent position $x = X(t)$—the shock being formed by crossing characteristics as shown in Figure 3.4. The conservation integral (3.12) must still hold even across such a shock. Let $x = a$ and $x = b$ be arbitrary points such that $a < X(t) < b$, let X^- denote a point just to the left of the shock, and X^+ denote a point just to the right, then (3.12) gives

$$\frac{\partial}{\partial t} \int_a^{X^-} \rho \, dx + \frac{\partial}{\partial t} \int_{X^+}^b \rho \, dx + q|_b - q|_{X^+} + q|_{X^+} - q|_{X^-} +$$

$$+q|_{X^-} - q|_a = \int_a^{X^-} r' dx + \int_{X^+}^b r' dx$$

$$\Rightarrow \int_a^{X^-} \frac{\partial \rho}{\partial t} dx + \dot{X} \rho|_{X^-} + \int_{X^+}^b \frac{\partial \rho}{\partial t} dx - \dot{X} \rho|_{X^+} + \int_a^{X^-} \frac{\partial q}{\partial x} dx +$$

$$+[q]_{X-}^{X+} + \int_{X+}^{b} \frac{\partial q}{\partial x} dx = \int_{a}^{X-} r' dx + \int_{X+}^{b} r' dx$$

$$\Rightarrow \int_{a}^{X-} \underbrace{\frac{\partial \rho}{\partial t} + \frac{\partial q}{\partial x} - r'}_{=0} dx + \int_{X+}^{b} \underbrace{\frac{\partial \rho}{\partial t} + \frac{\partial q}{\partial x} - r'}_{=0} dx -$$

$$-\dot{X} [\rho]_{X-}^{X+} + [q]_{X-}^{X+} = 0 \ .$$

Thus the shock moves according to the ordinary differential equation

$$\dot{X} = \frac{[q]}{[\rho]} \tag{3.13}$$

where [ ] denotes the jump in its argument across the shock, namely $[\ ]_{X-}^{X+}$. That is, the velocity of the shock= (jump in the flux)/(jump in the density).

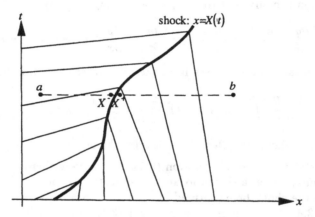

Figure 3.4: A shock formed between crossing characteristics in equation (3.11).

Such a solution of the differential equation (3.9), being smooth and differentiable everywhere except at a finite number of shocks where the shock relation (3.12) holds, is called a **weak solution** of the differential equation.

Example 3.1 Consider car traffic with a linear velocity-density relationship of $c(\rho) = 60(1 - 2\rho/150)$ km/hr. Then the flux-density relationship is parabolic: $q(\rho) = \int c(\rho)\,d\rho = 60\rho(1 - \rho/150)$ cars/hr. For the special initial condition

$$\rho_0(x) = \begin{cases} 25 & x < 0 \\ 75 & x > 0 \end{cases}$$

consisting of a shock at $x = 0$ separating a low-density uniform stream of fast cars ($v = q/\rho = 50$ km/hr) behind a medium-density uniform stream of slower cars ($v = q/\rho = 30$ km/hr). As seen in the characteristic diagram, Figure 3.5: the characteristics emanating from $t = 0$ for $x < 0$ have inverse-slope $\frac{dx}{dt} = c(25) = 40$ km/hr; while those characteristics emanating from $t = 0$ for $x > 0$ have inverse-slope $\frac{dx}{dt} = c(75) = 0$ km/hr. These would cross in the first quadrant and so an evolving shock must exist between the two uniform streams at all later times. Since the density ρ on each side of the shock is carried by the characteristics from the initial time $t = 0$, ρ must be a constant for each side of the shock, and so must the flux $q(\rho)$. Thus the speed of the shock is the constant

$$\dot{X} = \frac{[q]}{[\rho]} = \frac{2250 - 1250}{75 - 25} = 20 \text{ km/hr}.$$

The shock starts from $x = 0$ at $t = 0$ and is thus located at

$$x = X(t) = 20t \ .$$

Observe that individual cars, $v = 50$ km/hr, catch up to the shock, $v = 20$ km/hr, and then have to slow down suddenly to the velocity of the denser stream of cars, $v = 30$ km/hr.

Note: the shock relation $\dot{X} = [q]/[\rho]$ is easily obtainable from conservation principles in the shock region. In the case of car traffic, imagine an observer (a cyclist say) travelling with the shock and just behind it at $x = X^-(t)$, and another observer (cyclist) just in front at $x = X^+(t)$. The rate at which cars pass each cyclist has to be the same as otherwise cars would be piling up or be created in the shock region. Now the rate at which cars pass a cyclist= (rate past a fixed point) − (rate cyclist passes cars if they are fixed) $= q - \dot{X}\rho$. Thus

$$q_- - \dot{X}\rho_- = q_+ - \dot{X}\rho_+$$
$$\Rightarrow \dot{X}(\rho_+ - \rho_-) = q_+ - q_-$$
$$\Rightarrow \dot{X} = [q]/[\rho] \ .$$

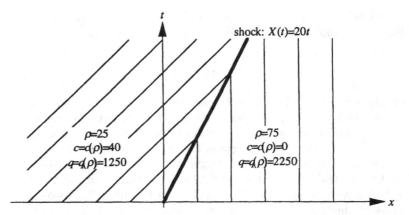

Figure 3.5: characteristic diagram of Example 3.1 showing the path of the shock between the two uniform streams down the road.

Example 3.2 In gas flow, shocks are frequently found at the extremes of what is called an N-wave, forming the characteristic double crack of a sonic boom. As described in Subsection 3.2.3 on loud unidirectional sound (in an ideal gas with $\gamma = 1$), a class of exact solutions to the dynamic gas equations may be found by restricting $\rho = \rho_* \exp(v/k)$, equation (3.7). Then the velocity field evolves according to equation (3.8), namely

$$\frac{\partial v}{\partial t} + (k+v)\frac{\partial v}{\partial x} = \frac{\partial v}{\partial t} + \frac{\partial q}{\partial x} = 0$$

where here the "flux" of the velocity is $q = kv + \frac{1}{2}v^2$. Consider the evolution following the N-wave initial condition shown at the bottom of Figure 3.6 and given algebraically by

$$v(x,0) = \begin{cases} 0 & |x| > \ell \\ x/\tau & |x| < \ell \end{cases},$$

where τ has the units of time and $\pm\ell/\tau$ are the extremes in the velocity v at the initial instant. The characteristic solution of the nonlinear equation (3.8) is

$$\frac{dv}{dt} = 0 \quad \text{on characteristics} \quad \frac{dx}{dt} = k + v .$$

As seen in the figure, the characteristics occupy three distinct regions: the undisturbed region in front of the N-wave, the N-wave region itself, and the region at rest behind the N-wave.

Algebraically, the front and the rear regions may be treated simultaneously as the initial condition, that $v = 0$, is identical in the two regions. Thus for the front and rear regions, the characteristic emanating from $x = s$ (for $|s| > \ell$) at $t = 0$ carries $v = 0$ and is consequently $x = kt + s$. Throughout the front and rear regions the velocity field is $v = 0$, that is, the gas is quiescent there. Within the N-wave, the characteristic emanating from $x = s$ (for $|s| < \ell$) at $t = 0$ carries $v = s/\tau$ and is consequently $x = (k + s/\tau)t + s$. As may be appreciated from the figure, the two types of characteristics would cross were it not for the the presence of the front and rear shock discontinuities.

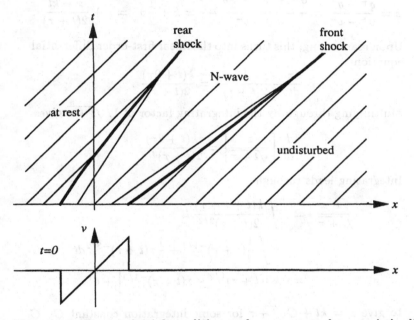

Figure 3.6: N-wave initial condition and consequent characteristic diagram in gas flow showing the spread of the wave.

Finding the locations of the shocks, $x(t)$, is a little harder in this example due to the spatial variations in v and the continuously varying slopes of the characteristics; however, the differential equation for their location is the same: $\dot{x} = [q]/[v]$. Suppose the rightmost shock (the one emanating from $x = \ell$) passes through a point (x, t). Then just to the right of the shock we will find an as yet undisturbed gas: $v^{+} = 0$ and hence $q^{+} = 0$. Just to the left of the shock are characteristics from the N-wave for which

$$x = (k + s/\tau)t + s = kt + (t/\tau + 1)s$$

$$\Rightarrow \quad s = \frac{x - kt}{t/\tau + 1} \quad \text{for } s \text{ as a function of } x \text{ and } t$$

$$\Rightarrow \quad v = s/\tau = \frac{x - kt}{\tau(t/\tau + 1)} = \frac{x - kt}{t + \tau} \ .$$

Thus the shock moves according to

$$\dot{x} = \frac{q^+ - q^-}{v^+ - v^-} = \frac{0 - \left(kv^- + \frac{1}{2}v^{-2}\right)}{0 - v^-} = k + \frac{1}{2}v^- = k + \frac{x - kt}{2(t + \tau)} \ .$$

Upon rearranging, this turns into the linear first-order differential equation

$$\frac{dx}{dt} - \frac{1}{2(t + \tau)}x = \frac{k(t + 2\tau)}{2(t + \tau)} \ .$$

Multiplying through by the integrating factor of $1/\sqrt{t + \tau}$ gives

$$\frac{d}{dt}\left[\frac{x}{\sqrt{t + \tau}}\right] = \frac{k(t + 2\tau)}{2(t + \tau)^{3/2}}$$

Integrating leads through

$$\begin{aligned}
\frac{x}{\sqrt{t + \tau}} &= \int \frac{k(t + \tau + \tau)}{2(t + \tau)^{3/2}}dt \\
&= k\int \frac{1}{2}(t + \tau)^{-1/2} + \frac{1}{2}\tau(t + \tau)^{-3/2}dt \\
&= k\left[(t + \tau)^{1/2} - \tau(t + \tau)^{-1/2}\right] + C
\end{aligned}$$

to give $x = kt + C\sqrt{t + \tau}$ for some integration constant C. C is found from the initial condition that at $t = 0$ the front shock starts at $x = \ell$, hence $\ell = 0 + C\sqrt{\tau}$ giving $C = \ell/\sqrt{\tau}$. Thus the location of the front shock is

$$x = kt + \ell\sqrt{1 + t/\tau} \ ,$$

and similarly it may be shown that the rear shock is located at $x = kt - \ell\sqrt{1 + t/\tau}$.

As shown schematically in the figure, the resultant N-wave between the two shocks is centred on the line $x = kt$ and has a width, $2\ell\sqrt{1+t/\tau}$, which increases like the square root of time. The velocity structure in the N-wave remains linear for all time as can be seen in the equation $v = \frac{x-kt}{t+\tau}$. The overall strength of the N-wave can be measured by the maximum velocity of the air in the wave, v_m, which is initially $v_m(0) = \ell/\tau$. At later times the maximum gas velocity is found just behind the front shock, namely v^-. Thus

$$v_m(t) = \frac{x-kt}{t+\tau} = \frac{\ell\sqrt{1+t/\tau}}{\tau(1+t/\tau)} = \frac{v_m(0)}{\sqrt{1+t/\tau}} \ ,$$

and so for large time the maximum velocity in the N-wave decays like one on the square root of time.

This decay and spread of the N-wave is characteristic of shock waves in air. A complicated initial condition for the gas flow, for example the multiplicity of shocks generated on the leading and trailing edges of a supersonic aeroplane, typically evolves through weakening and merger into such an N-wave. The ubiquitous presence of the factor $(1+t/\tau)$ in the time dependence indicates that this N-wave has a virtual origin at a time $t = -\tau$ when it was infinitesimally thin, but infinitely strong! Incidentally, the shock wave from a supersonic aeroplane decays somewhat faster than $t^{-1/2}$ due to the additional effect of the shock wave spreading out in two dimensions—the above analysis is just for one-dimensional propagation.

3.3.2 Characteristics for higher-order PDEs

So far we have used the method of characteristics to solve exactly only a *single first-order* differential equations. When anything more complicated is derived as the mathematical model, such as the two coupled differential equations (3.4) and (3.5) for gas flow, we either linearised the equations (Subsection 3.2.2) or investigated only that restricted class of solutions which can be described by one first-order differential equation (Subsection 3.2.3). For ease of exposition I will continue to do this throughout most of the book. However, it is useful to investigate briefly how the method of characteristics may be extended to deal with almost arbitrarily complicated nonlinear differential equations. This is the topic of this subsection.

Consider the general system of n nonlinear differential equations

$$A\frac{\partial u}{\partial t} + B\frac{\partial u}{\partial x} = d \tag{3.14}$$

for the n-vector of unknowns $u(x,t)$, a function of one space variable x and time t. The coefficients in this equation, the $n \times n$ matrices $\mathcal{A}$ and $\mathcal{B}$ and the n-vector d, may be functions of the unknown u and of x and t, but may not depend upon the derivatives of u. This forms a set of n **quasi-linear partial differential equations**—the term quasi-linear refers to the linearity of the equations in the derivatives of u, but allowing a nonlinear dependence upon u itself.

Many physically based equations may be put into this form.

Example 3.3 The equations (3.4) and (3.5) for the dynamics of an ideal gas (with $F = 0$) may be written as

$$
\begin{array}{llll}
\frac{\partial \rho}{\partial t} & +v\frac{\partial \rho}{\partial x} & +\rho\frac{\partial v}{\partial x} & = 0 \\
+\rho\frac{\partial v}{\partial t} & +k^2\rho^{\gamma-1}\frac{\partial \rho}{\partial x} & +\rho v\frac{\partial v}{\partial x} & = 0
\end{array}
$$

which with $u = \begin{bmatrix} \rho \\ v \end{bmatrix}$ becomes

$$
\underbrace{\begin{bmatrix} 1 & 0 \\ 0 & \rho \end{bmatrix}}_{\mathcal{A}} \frac{\partial u}{\partial t} + \underbrace{\begin{bmatrix} v & \rho \\ k^2\rho^{\gamma-1} & \rho v \end{bmatrix}}_{\mathcal{B}} \frac{\partial u}{\partial x} = \underbrace{\mathbf{0}}_{d}
$$

Example 3.4 As is discussed in Section 5.3, the flow of water in a channel or river may be adequately described by a one-dimensional model. In this case if we let $h(x,t)$ be the depth of the water and $v(x,t)$ the average velocity of the water, then conservation of water gives $\frac{\partial h}{\partial t} + \frac{\partial(hv)}{\partial x} = 0$, equation (5.7), while conservation of momentum gives $\frac{\partial v}{\partial t} + v\frac{\partial v}{\partial x} + g\frac{\partial h}{\partial x} = 0$, equation (5.8). These may be immediately written as

$$
\begin{array}{llll}
\frac{\partial h}{\partial t} & +v\frac{\partial h}{\partial x} & +h\frac{\partial v}{\partial x} & = 0 \\
+\frac{\partial v}{\partial t} & +g\frac{\partial h}{\partial x} & +v\frac{\partial v}{\partial x} & = 0
\end{array}
$$

which with $u = \begin{bmatrix} h \\ v \end{bmatrix}$ becomes

$$
\underbrace{\begin{bmatrix} 1 & 0 \\ 0 & 1 \end{bmatrix}}_{\mathcal{A}} \frac{\partial u}{\partial t} + \underbrace{\begin{bmatrix} v & h \\ g & v \end{bmatrix}}_{\mathcal{B}} \frac{\partial u}{\partial x} = \underbrace{\mathbf{0}}_{d}
$$

Characteristic solution

A general system of n nonlinear partial differential equations such as (3.14) has n characteristic directions through every point (x, t). These n local directions must then be assembled by integration to form n families of characteristic curves. There is always a good physical interpretation of the n families of characteristics. In the case of an ideal gas, where $n = 2$, one family of characteristics corresponds to left travelling sound waves while the other family describes right travelling waves.

We seek n characteristic curves satisfying an ordinary differential equation of the form

$$\frac{dx}{dt} = \lambda(u, x, t) \tag{3.15}$$

and on which the partial differential equation (3.14) reduces to an ordinary differential equation (for example, when $n = 1$ we would aim to derive $\frac{du}{dt} = r(u, x, t)$). To do this solve (at every point (x, t)!) the generalised eigen-problem

$$\boxed{z^{\mathsf{T}} \mathcal{B} = \lambda z^{\mathsf{T}} \mathcal{A},} \tag{3.16}$$

for eigenvalues λ and left-eigenvectors z^{T}. That is, solving the characteristic equation $|\mathcal{B} - \lambda\mathcal{A}| = 0$ gives the eigenvalues, and then solving the linear equations $z^{\mathsf{T}}[\mathcal{B} - \lambda\mathcal{A}] = o$ gives the corresponding left-eigenvectors. If this supplies n real and distinct eigenvalues λ_j (and corresponding real eigenvectors z_j) then the partial differential equation is purely **hyperbolic**. If some of the eigenvalues are complex then the equation has a mixed nature, partly elliptic and partly hyperbolic, while if all the eigenvalues are complex then the equation is purely elliptical.

Example 3.5 Consider a general single second-order partial differential equation

$$A\frac{\partial^2 u}{\partial x^2} + B\frac{\partial^2 u}{\partial x \partial t} + C\frac{\partial^2 u}{\partial t^2} = D$$

where the coefficients A, B, C and D may be functions of x, t, $\frac{\partial u}{\partial x}$ and $\frac{\partial u}{\partial t}$. This is put in the form (3.14) by letting $u_1 = \frac{\partial u}{\partial x}$ and $u_2 = \frac{\partial u}{\partial t}$. Then, since $\frac{\partial u_1}{\partial t} = \frac{\partial u_2}{\partial x}$,

$$A\frac{\partial u_1}{\partial x} \quad +B\frac{\partial u_2}{\partial x} \qquad +C\frac{\partial u_2}{\partial t} \quad = D$$
$$\qquad\qquad -\frac{\partial u_2}{\partial x} \qquad +\frac{\partial u_1}{\partial t} \qquad = 0$$

$$\Rightarrow \quad \underbrace{\begin{bmatrix} A & B \\ 0 & -1 \end{bmatrix}}_{\mathcal{B}} \frac{\partial u}{\partial x} + \underbrace{\begin{bmatrix} 0 & C \\ 1 & 0 \end{bmatrix}}_{\mathcal{A}} \frac{\partial u}{\partial t} = \underbrace{\begin{bmatrix} D \\ 0 \end{bmatrix}}_{d}$$

Here

$$|\mathcal{B} - \lambda \mathcal{A}| = -A + \lambda(B - \lambda C) = -\left[A - B\lambda + C\lambda^2\right] = 0 \, ,$$

which using $\frac{dx}{dt} = \lambda$ is

$$A - B\left(\frac{dx}{dt}\right) + C\left(\frac{dx}{dt}\right)^2 = 0 \, .$$

If $B^2 - 4AC > 0$ then this quadratic has two real roots giving two real characteristics and the partial differential equation is hyperbolic. If $B^2 - 4AC < 0$ then the two roots are complex and the equation is elliptic. The case $B^2 - 4AC = 0$ is the degenerate case when only one characteristic can be found and the equation is called parabolic.

It is only in the purely hyperbolic case that the method of characteristics gives a complete method of solution to the problem. Then, for each $j = 1, \ldots, n$ the scalar quantity

$$\begin{aligned}
z_j^\mathsf{T} d &= z_j^\mathsf{T} \left(\mathcal{A} u_t + \mathcal{B} u_x\right) \\
&= z_j^\mathsf{T} \mathcal{A} u_t + z_j^\mathsf{T} \mathcal{B} u_x \\
&= z_j^\mathsf{T} \mathcal{A} u_t + \lambda_j z_j^\mathsf{T} \mathcal{A} u_x \quad \text{by (3.16)} \\
&= z_j^\mathsf{T} \mathcal{A} \left(u_t + \lambda_j u_x\right) \\
&= z_j^\mathsf{T} \mathcal{A} \frac{du}{dt} \quad \text{on curves } \frac{dx}{dt} = \lambda_j \text{ by chain rule}
\end{aligned}$$

Letting $\zeta_j^\mathsf{T} = z_j^\mathsf{T} \mathcal{A}$ for brevity this gives

$$\boxed{\zeta_j^\mathsf{T} \frac{du}{dt} = \left(z_j^\mathsf{T} d\right) \quad \text{on characteristics } \mathcal{C}_j \quad \frac{dx}{dt} = \lambda_j} \qquad (3.17)$$

Note that differentiation along the jth characteristic $\mathcal{C}_j$ is implicit in the derivative $\frac{du}{dt}$.

The role of equation (3.17) in the characteristic solution is hard to appreciate. To see how a solution field is constructed consider the case of a pair $(n = 2)$ of quasi-linear partial differential equations and imagine constructing a simple numerical scheme based on (3.17). Identifying the components of $\zeta_j^T = (\alpha_j, \beta_j)$ and writing $\gamma_j = z_j^T d$, equation (3.17) reads

$$\alpha_j \frac{du_1}{dt} + \beta_j \frac{du_2}{dt} = \gamma_j \quad \text{on } \mathcal{C}_j \quad \frac{dx}{dt} = \lambda_j \quad \text{for } j = 1, 2.$$

As shown in Figure 3.7, given a point (x', t') with solution value u', the two characteristics propagate information in two different directions, namely $\mathcal{C}_1'$ and $\mathcal{C}_2'$. At the slightly later time t the $\mathcal{C}_1'$ characteristic has sent the information that $\alpha_1'(u_1 - u_1') + \beta_1'(u_2 - u_2') = \gamma_1'(t - t')$ to a point (x, t). However, there are two unknowns and this is only one equation for the unknowns; hence we need more information. As seen in the figure, this comes from the $\mathcal{C}_2$ characteristic passing through the point (x, t) from another point earlier in time (x'', t''). This provides the equation $\alpha_2''(u_1 - u_1'') + \beta_2''(u_2 - u_2'') = \gamma_2''(t - t'')$.

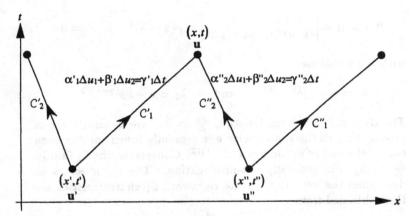

Figure 3.7: the diagram showing the basis of a simple numerical scheme for using characteristics to solve a pair of coupled quasi-linear PDEs.

Information about the change in u_1 and u_2 along both $\mathcal{C}_1'$ and $\mathcal{C}_2''$ is needed to determine u at (x, t). This reflects the property that both left and right travelling waves contribute to the solution at any point.

Riemann invariants

If $d = o$ in the quasi-linear partial differential equation then the subsequent analysis simplifies (just as in car traffic where the characteristics become straight lines on which the density is constant). In this case the characteristic solution is $\zeta_j^{\mathsf{T}} \frac{du}{dt} = 0$ on characteristics $\mathcal{C}_j \frac{dx}{dt} = \lambda_j$. That is, along characteristics the linear combinations $\zeta_j^{\mathsf{T}} \frac{du}{dt}$ do not change. In conjunction with the property that most physically based equations are homogeneous in space and time (that is, independent of location), this often allows the differential equation along the characteristics to be integrated. Integration then shows that certain combinations of the unknown u are constant, called the **Riemann invariants**. This is best seen in an example.

Example 3.6 Consider the equations for ideal gas dynamics. To put them into the framework of the general theory we earlier identified $u = \begin{bmatrix} \rho \\ v \end{bmatrix}$, $\mathcal{A} = \begin{bmatrix} 1 & 0 \\ 0 & \rho \end{bmatrix}$, $\mathcal{B} = \begin{bmatrix} v & \rho \\ k^2\rho^{\gamma-1} & \rho v \end{bmatrix}$, and $d = o$. Then the characteristic equation becomes

$$|\mathcal{B} - \lambda\mathcal{A}| = \begin{vmatrix} v - \lambda & \rho \\ k^2\rho^{\gamma-1} & \rho v - \rho\lambda \end{vmatrix} = \rho(v - \lambda)^2 - k^2\rho^\gamma = 0 \ ,$$

with the solutions

$$\lambda_1 = v + k\rho^{(\gamma-1)/2} \qquad \text{and} \qquad \lambda_2 = v - k\rho^{(\gamma-1)/2} \ .$$

The $\mathcal{C}_1$ family of characteristics, $\frac{dx}{dt} = \lambda_1$, are generally right-propagating as the gas velocity v is generally much smaller than the local speed of sound $c = k\rho^{(\gamma-1)/2}$. Conversely, the $\mathcal{C}_2$ family, $\frac{dx}{dt} = \lambda_2$, are generally left-propagating. The eigenvectors to determine the evolution on the rightward $\mathcal{C}_1$ characteristics are then obtained from

$$z_1^{\mathsf{T}} (\mathcal{B} - \lambda\mathcal{A}) = o^{\mathsf{T}}$$

$$\Rightarrow \ z_1^{\mathsf{T}} \begin{bmatrix} -k\rho^{(\gamma-1)/2} & \rho \\ k^2\rho^{\gamma-1} & -k\rho^{(\gamma+1)/2} \end{bmatrix} = o^{\mathsf{T}}$$

$$\Rightarrow \ z_1^{\mathsf{T}} = \left(k\rho^{(\gamma-1)/2}, 1 \right)$$

$$\Rightarrow \ \zeta_1^{\mathsf{T}} = z_1^{\mathsf{T}} \mathcal{A} = \left(k\rho^{(\gamma-1)/2}, \rho \right)$$

Thus on each $\mathcal{C}_1$ characteristic

$$\zeta_1^{\mathsf{T}} \frac{du}{dt} = k\rho^{(\gamma-1)/2} \frac{d\rho}{dt} + \rho \frac{dv}{dt} = 0 \ .$$

In this case, as in many others, this last equation can be integrated. Here, this is achieved by multiplying through by the integrating factor $1/\rho$ to give

$$k\rho^{(\gamma-3)/2}\frac{d\rho}{dt} + \frac{dv}{dt} = \frac{d}{dt}\left(\frac{2k}{\gamma-1}\rho^{(\gamma-1)/2} + v\right) = 0$$

and hence $\frac{2k}{\gamma-1}\rho^{(\gamma-1)/2} + v = $ constant. This constant is in general different for each C_1 characteristic as the integration is done along each such characteristic independent of its neighbours.

Similarly, on each C_2 characteristic we can derive that $-\frac{2k}{\gamma-1}\rho^{(\gamma-1)/2} + v = $ constant. The combinations

$$\pm\frac{2k}{\gamma-1}\rho^{(\gamma-1)/2} + v$$

are called **Riemann invariants**.

Example 3.7 Consider the propagation of sound into still air. As in Problem 3.5, we imagine a semi-infinite pipe $0 < x < \infty$ filled with an ideal gas which is initially at rest, $v = 0$ and $\rho = \rho_0$ at $t = 0$. An ideal fan located at $x = 0$ blows in or sucks out gas with some prescribed velocity $v_0(t)$ which is never faster than the speed of sound. The characteristic solution, that

$$v \pm \frac{2k}{\gamma-1}\rho^{(\gamma-1)/2} = \text{constant on lines} \quad \frac{dx}{dt} = v \pm k\rho^{(\gamma-1)/2}$$

leads to a characteristic diagram such as the one depicted in Figure 3.8.

In the *still region*: each C_1 characteristic emanates from still gas, $\rho = \rho_0$ and $v = 0$, which implies the corresponding Riemann invariant $v + \frac{2k}{\gamma-1}\rho^{(\gamma-1)/2} = \frac{2k}{\gamma-1}\rho_0^{(\gamma-1)/2}$; and each C_2 characteristic also emanates from still gas which implies $v - \frac{2k}{\gamma-1}\rho^{(\gamma-1)/2} = -\frac{2k}{\gamma-1}\rho_0^{(\gamma-1)/2}$. These form a pair of simultaneous equations for ρ and v which have the solution

$$\rho = \rho_0 \quad \text{and} \quad v = 0$$

everywhere in the still region.

In the *disturbed region*: the relation

$$v - \frac{2k}{\gamma-1}\rho^{(\gamma-1)/2} = -\frac{2k}{\gamma-1}\rho_0^{(\gamma-1)/2}$$

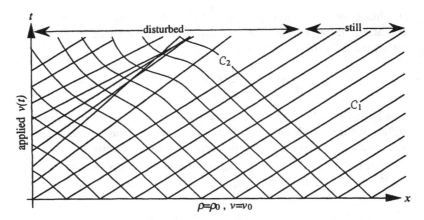

Figure 3.8: possible characteristic diagram for a fan blowing into a tube of still, ideal gas.

is carried in by all the C_2 characteristics. On the other hand, each C_1 characteristic now emanates from the fan and so has

$$v + \frac{2k}{\gamma - 1}\rho^{(\gamma-1)/2} = K$$

where K is a constant along each C_1 characteristic but will typically vary from one C_1 characteristic to another. The particular value of K on a particular C_1 is determined from the behaviour of the fan at the instant the characteristic leaves the fan. These two equations can once again be solved to obtain the constant values of v and ρ on each C_1 characteristic, it is just that their values change from one characteristic to another. Since v and ρ are constant on each C_1 characteristic it follows that each C_1 characteristic has constant slope, $\frac{dx}{dt} = v + k\rho^{(\gamma-1)/2}$, and is therefore a straight line (as shown in the figure). However, the C_2 characteristics in the disturbed region are not straight as they are crossing C_1 characteristics possessing different combinations of v and ρ.

ASIDE   As in the earlier simpler studies, this example shows that most fan motions generate crossing characteristics in the one family, which here is the C_1 family, and so, in the absence of dissipation, a "shock" will form to give a sonic boom. The other point of interest is that the earlier analysis of loud uni-directional sound (Section 3.2.3) gives precisely the same equations as obtained here for the C_1 characteristics when they propagate into an undisturbed region of gas. Thus the assumption made there of having a functional relation between the velocity v and the density ρ is precisely equivalent to assuming propagation into an undisturbed region!

Exercises

Problem 3.6   Water flowing down a sloping rough river bed travels   EASY with a velocity $v = \sqrt{g\alpha/C}h^{1/2}$ where h is the depth of the water, g is the acceleration of gravity, α is the slope of the bed, and C is a roughness coefficient. This relation between water depth and velocity is called the **Chezy law**, see Subsection 5.3.2. Conservation of water additionally provides the dynamical equation $\frac{\partial h}{\partial t} + \frac{\partial(hv)}{\partial x} = 0$, see Section 5.3.

(a) A flash flood roars down a dry river bed. Use the jump conditions to determine the velocity of the "wall of water" at the front of the flood as it travels downstream. If the river bed in front of the flood is not dry, but has some water in it, is the velocity of the "wall" faster or slower than that for the dry river bed?

(b) A deluge in the hills supplies a significant increase, albeit fairly short-lived, in the flux of water entering a river plain at the foothills at $x = 0$. By sketching the characteristic diagram, show that this model always predicts that a "shock" will always form somewhere downstream of the foothills.

Problem 3.7   Consider the ideal gas equations for $\gamma = 1$. Find the   EASY characteristic solution of the equations and then derive the Riemann invariants. Assuming that right-travelling waves are propagating into an undisturbed region, show that this characteristic solution reduces to the unidirectional sound solutions of Section 3.2.3.

Problem 3.8   Consider the pair of quasi-linear equations for long   MEDIUM water waves given in Example 3.4.

(a) Show that the characteristic solution gives $v \pm 2\sqrt{gh} = $ constant (the Riemann invariants) on characteristics $\frac{dx}{dt} = v \pm \sqrt{gh}$.

(b) Picture a semi-infinite tank $0 < x < \infty$ of still water of depth h_*. For $t > 0$ a wavemaker at $x = 0$ applies a given velocity $v(t)$ to the water at $x = 0$. Draw a typical characteristic diagram for this situation and discuss the features shown by the characteristic solution.

MEDIUM   **Problem 3.9**   Consider the pair of quasi-linear equations for blood flow in inactive arteries as given by Equations (5.4–5.5). Show that the characteristic solution gives $v \pm 2\sqrt{\frac{2\alpha R}{\rho}} = $ constant (the Riemann invariants) on characteristics $\frac{dx}{dt} = v \pm \sqrt{\frac{\alpha R}{2\rho}}$.

ASIDE   However, note that for this nonlinear solution to be physically relevant we should first have modelled the elastic behaviour of the artery wall by an appropriate nonlinear model rather than assume the linear Hookean relation between pressure and artery stretching.

Chapter 4

Stress and strain

In a general continuum, as seen in Section 3.2 on ideal gases, we have
two differential equations, equations (2.1) and (3.2), for the three unknown fields ρ, v and σ. This is apparently two-thirds of the work
already done. However, it is the remaining third which is the prime
difficulty: namely, relating the pattern of deformation of the material
to the resultant pattern of stress.

Loosely speaking:

1. **stress** is the force applied by one piece of material upon a neighbouring piece, the pattern of internal stress affects the movement
   of the material ($\frac{\partial \sigma}{\partial x}$ in the momentum equation);

2. **strain** measures how much the material has deformed from a reference state, and hence may be found from any known deformation;

3. stress and strain should clearly be related by an equation, like
   the equation of state for an ideal gas, which completes the set of
   equations.

4.1   Measures of strain

Our first task is to define how to describe the **strain**: namely that part
of a material's deformation which produces the internal stresses. Initially we distinguish between an Eulerian description and a Lagrangian
description of the deformation of a continuum; however, we observe that

77

for practical purposes there is little difference between the two systems. Thus at the end of this section we return to an Eulerian description.

4.1.1   Uniform straining

Consider a bar of material of some initial length ℓ_0. If in some deformation it is stretched to a length ℓ then it is natural to describe the change by the **strain** which is defined to measure its proportional change in length:

$$\text{either} \quad \tilde{e}^L = \frac{\ell - \ell_0}{\ell_0} \quad \text{or} \quad \tilde{e}^E = \frac{\ell - \ell_0}{\ell}$$

are used, depending upon convenience. Numerically they are different: if $\ell_0 = 1$ and the bar is stretched to $\ell = 2$ then $\tilde{e}^L = 1$ but $\tilde{e}^E = \frac{1}{2}$; while if it is compressed to $\ell = \frac{1}{2}$ then $\tilde{e}^L = -\frac{1}{2}$ but $\tilde{e}^E = -1$. However, a positive strain is always an extension, while a negative strain is always a contraction.

ASIDE

Note: the symbol e for extension is the main symbol to denote a strain. The superscript ^{L} is used above because there the reference length is the *original* length ℓ_0, and so it is like a Lagrangian quantity. The superscript ^{E} is used because there the reference length is the *current* length ℓ, and so it is like an Eulerian quantity.

Actually, it is more convenient to use one of the measures of strain

$$e^L = \frac{\ell^2 - \ell_0^2}{2\ell_0^2} \quad \text{or} \quad e^E = \frac{\ell^2 - \ell_0^2}{2\ell^2} \ .$$

For example, if $\ell_0 = 1$ and $\ell = 2$, as before, then $e^L = \frac{3}{2}$ but $e^E = \frac{3}{8}$. However, if a bar of initial length $\ell_0 = 1.00$ is slightly stretched to $\ell = 1.01$ then

$$
\begin{aligned}
\tilde{e}^L &= 0.01 \\
\tilde{e}^E &= 0.00990 \approx 0.01 \\
e^L &= 0.01005 \approx 0.01 \\
e^E &= 0.00985 \approx 0.01 \ .
\end{aligned}
$$

In small elongations, all of the above strain measures are effectively equal.

For a bar of real material, like metal, the stress is related to the strain as shown in Figure 4.1. Such a figure would be obtained by placing a rod of the material into a testing machine and applying a force (stress)

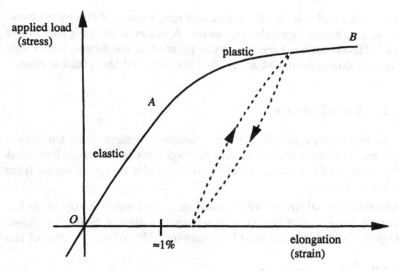

Figure 4.1: a typical stress–strain relationship (applied force–elongation relationship) of a bar of a material placed in a testing machine

to each end and then measuring the resultant deformation (strain). To a good approximation the region OA on the figure is linear and hence

$$\sigma \approx E \tilde{e}^L$$

where E is a constant of the material called **Young's modulus.**

ASIDE

The dashed lines in Figure 4.1 are included to represent the fact that a metal bar can be deformed so much that it has enters the *plastic* regime AB, after which it does not return to its original length. Instead, if the applied stress is reduced then the elongation of the rod follows the dashed lines rather than the original solid line of elastic behaviour. In some materials this phenomenon is termed *work-hardening*, and may provide a useful way to increase the maximum stress which the material can support and thus increase its strength.

The above linear relation between stress and strain is a good approximation for strains less than that corresponding to the point A on the figure. However, the point A typically corresponds to strains in the region of 1% for which, as we have already noted, $\tilde{e}^L \approx \tilde{e}^E \approx e^L \approx e^E$. Hence we write **Hooke's law,** that

$$\sigma = Ee ,   \tag{4.1}$$

where e is an **infinitesimal strain** and may be any of the above measures, or a different equivalent measure. A material obeying this law is
called a **Hookean material**, and the point A in the figure, where this
law breaks down, occurs at an applied stress called the **yield stress**.

4.1.2   Local strain

The previous discussion is fine for a uniform deformation, the strain
and stress then both being constant throughout the material. But what
is the local strain when the deformation of the material varies from
location to location?

Consider an arbitrary deformation described equivalently by either
$x = x^L(\xi, t)$ or $\xi = \xi^E(x, t)$, and fix upon a time $t > 0$. As shown
in Figure 4.2, an infinitesimal line element PP', of length $d\ell_0$, of the

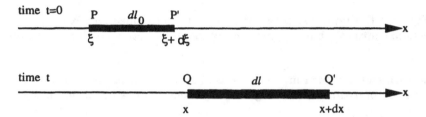

Figure 4.2: a diagram of the deformation of a small part of a bar.

material at time $t = 0$ will at time t have been moved to the line element
QQ', of length $d\ell$, by the deformation. Clearly

$$d\ell = dx = \frac{\partial x^L}{\partial \xi} d\xi \quad \text{and} \quad d\ell_0 = d\xi = \frac{\partial \xi^E}{\partial x} dx \ .$$

Thus the difference in the squares of the length from after to before the
deformation is

$$d\ell^2 - d\ell_0^2 \;=\; dx^2 - d\xi^2 = \left[1 - \left(\frac{\partial \xi^E}{\partial x}\right)^2\right] dx^2 \quad \text{(Eulerian)}$$

$$d\ell^2 - d\ell_0^2 \;=\; dx^2 - d\xi^2 = \left[\left(\frac{\partial x^L}{\partial \xi}\right)^2 - 1\right] d\xi^2 \quad \text{(Lagrangian)}$$

Hence, how much each small line element is stretched is measured either by **Almansi's strain**

$$e^E(x,t) = \frac{d\ell^2 - d\ell_0^2}{2d\ell^2} = \frac{1}{2}\left[1 - \left(\frac{\partial\xi^E}{\partial x}\right)^2\right]$$

or by **Green's strain**

$$e^L(\xi,t) = \frac{d\ell^2 - d\ell_0^2}{2d\ell_0^2} = \frac{1}{2}\left[\left(\frac{\partial x^L}{\partial\xi}\right)^2 - 1\right],$$

just as for the case of uniform stretching but now for each little segment of the material.

Example 4.1 Consider the deformation $x^L = \xi + \xi t^2$ or equivalently $\xi^E = x/(1+t^2)$ as used in Examples 1.3 and 1.4. For this deformation

$$e^L = \frac{1}{2}\left[(1+t^2)^2 - 1\right] = t^2 + \frac{1}{2}t^4$$

while

$$e^E = \frac{1}{2}\left[1 - \frac{1}{(1+t^2)^2}\right] = \frac{t^2(1+\frac{1}{2}t^2)}{(1+t^2)^2}.$$

It is a coincidence that the strain happens to be the same for all particles at any time. However, it is always true that for small deformations $e^L \approx e^E$. Here small deformations occur for small times when we observe that $e^L \approx t^2 \approx e^E$.

Define the **displacement field** to be $u = x - \xi$ which is simply the distance between the current position of a particle and its initial position. Hence introduce the two new functions: $u^L(\xi,t) = x^L(\xi,t) - \xi$, the Lagrangian description of the displacement; and $u^E(x,t) = x - \xi^E(x,t)$, the Eulerian description of the displacement. In terms of these functions the strains become

$$\begin{aligned}
e^E &= \frac{1}{2}\left[1 - \left(\frac{\partial\xi^E}{\partial x}\right)^2\right] \\
&= \frac{1}{2}\left[1 - \left(1 - \frac{\partial u^E}{\partial x}\right)^2\right] \quad \text{since } \xi^E = x - u^E \\
&= \frac{\partial u^E}{\partial x} - \frac{1}{2}\left(\frac{\partial u^E}{\partial x}\right)^2
\end{aligned}$$

$$\text{and} \quad e^L = \frac{1}{2}\left[\left(\frac{\partial x^L}{\partial \xi}\right)^2 - 1\right]$$

$$= \frac{1}{2}\left[\left(1 + \frac{\partial u^L}{\partial \xi}\right)^2 - 1\right] \quad \text{since } x^L = \xi + u^L$$

$$= \frac{\partial u^L}{\partial \xi} + \frac{1}{2}\left(\frac{\partial u^L}{\partial \xi}\right)^2 .$$

The point of introducing the displacement u is this: if the deformation is nearly a rigid-body one (as is typical, as strains are usually less than 1%) then the displacement u is nearly constant, which implies that $\frac{\partial u^E}{\partial x}$ is small and that $\left(\frac{\partial u^E}{\partial x}\right)^2$ is very small and negligible compared to the dominant term $\frac{\partial u^E}{\partial x}$. Thus the expression for Almansi's strain e^E is effectively the same as

$$e = \frac{\partial u}{\partial x} , \tag{4.2}$$

which is called **Cauchy's infinitesimal strain**. Similarly, $e^L \approx \frac{\partial u}{\partial \xi}$. But, since the deformation is nearly a rigid-body one, $\frac{\partial u}{\partial \xi} \approx \frac{\partial u}{\partial x}$ and therefore Green's strain e^L is also effectively the same as Cauchy's infinitesimal strain e.

In the common situation of nearly rigid-body deformations, the distinction between the various measures of strain disappears—they all measure the local stretching of the material. We generally use Cauchy's infinitesimal strain as it has the simplest definition.

Exercises

EASY

Problem 4.1 Calculate Almansi's strain e^E and Green's strain e^L for all time for the deformations:

$$\text{(i)} \quad x = \xi + v_0 t + \xi t^2 ; \qquad \text{(ii)} \quad x = \xi + \frac{1}{2}t\xi^2 .$$

Also calculate the displacement functions $u^L(\xi, t)$ and $u^E(x, t)$, and then verify that for small times (that is, small deformations) $e^E \approx e^L \approx \frac{\partial u^E}{\partial x} \approx \frac{\partial u^L}{\partial \xi}$.

EASY

Problem 4.2 Consider an arbitrary, but small, uniform stretching of a rod from a length ℓ_0 to a length $\ell = \ell_0 + \delta\ell$. Find the two term Taylor series in $\delta\ell$ for all four strain measures, $\tilde{e}^L$, $\tilde{e}^E$, e^L and e^E. By comparing their linear dependence upon $\delta\ell$ show that they are effectively the same for this small deformation.

Problem 4.3 It is easy to establish that $\frac{\partial u^L}{\partial t} = v^L$; do so. However, MEDIUM it is more useful to establish the relationship between displacement and velocity in an Eulerian description of a deformation. Argue that

$$\frac{Du^E}{Dt} = \frac{\partial u^E}{\partial t} + v^E \frac{\partial u^E}{\partial x} = v^E . \qquad (4.3)$$

Hint: the answer to Problem 1.5 may help. Verify this relation for the deformation used in Problem 1.3.

> We now return to using our default convention of using an Eulerian   ASIDE
> description of a deformation, and so we drop the indicative superscript.

4.2   Elastic deformation of a bar

We now turn our attention to the behaviour of a beam or rod of some solid material. Their behaviour is very important as many engineering structures are based on a framework of beams.

The equations we base our predictions on are: the continuity equation (2.1); the momentum equation (3.2); the stress-strain relationship given by Hooke's law (4.1); Cauchy's definition of infinitesimal strain (4.2); and the relation between displacement and velocity, equation (4.3). These equations are to be solved for the five unknowns: ρ, v, σ, e and u. This sounds a daunting task. Fortunately, as already noted in the previous section, typical deformations of a bar are small (less than 1%) and we linearise all these equations immediately; this makes many of them nearly trivial.

> Note that Hooke's law was originally intended to apply only to the   ASIDE
> stretching of a rod as a whole. However, we are going to use the law as
> if it applies to every small segment of the rod individually.

4.2.1   p-waves

Consider a rod of some material and imagine giving it a push on one end. How is this push transmitted through the rod? On a human-scale this push is transmitted almost instantaneously—there is no detectable delay in pushing one end of a ruler and the other end starting to move. However, on a larger scale the delay in the transmission is evident; for example, in the propagation of earthquake waves away from the quake and through the earth.

Consider an undeformed rod which is at rest, is of constant density ρ_*, and is not subject to any body forces. We linearise the governing equations by looking at the behaviour of "small" disturbances to this state. Thus let $\rho = \rho_* + \hat{\rho}(x,t)$ and consider that the density variations $\hat{\rho}$, the velocity v, the displacement u, the strain e and the stress σ are all "small" quantities. Starting with the linearised momentum equation (3.2)

$$\rho_* \frac{\partial v}{\partial t} \approx \frac{\partial \sigma}{\partial x}$$

$$\approx E\frac{\partial e}{\partial x} \quad \text{by Hooke's law (4.1)}$$

$$\approx E\frac{\partial^2 u}{\partial x^2} \quad \text{by the definition of strain (4.2) .}$$

But also equation (4.3) shows that $v \approx \frac{\partial u}{\partial t}$ and so the above equation becomes, for "small" displacements,

$$\frac{\partial^2 u}{\partial t^2} = \frac{E}{\rho_*} \frac{\partial^2 u}{\partial x^2} \; . \tag{4.4}$$

This, once again, is the wave equation describing the propagation of disturbances, unchanged in form, with a wave-speed $c = \sqrt{E/\rho_*}$. It describes the propagation of sound waves along the rod.

Observe that the continuity equation (2.1) was not used in the derivation of this wave equation. It just so happens that for waves in a Hookean solid, the continuity equation has no significant dynamic role to play—it just governs how the density field responds to the displacement fluctuations.

We shall return to these p-waves later, in Section 4.3, where we investigate the effects on a bar governed by more realistic relations between stress and strain than that given by Hooke's law (4.1).

4.2.2   The bending of a beam

A fundamental problem in the design of structures is to find how much a beam bends when it is loaded with a sideways force $w(x)$, as shown in

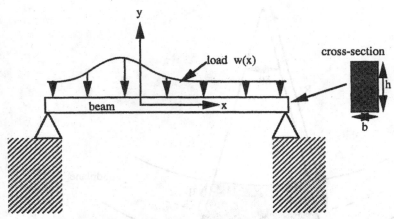

Figure 4.3: A diagram of a beam with a loading $w(x)$

Figure 4.3. It is not strictly a one-dimensional phenomenon because the sideways bending, as described by the unknown deflection $y = Y(x)$, is clearly in a direction at right-angles to the rod. Nonetheless, it is the distribution of various forces and moments along the beam which is vital, and it is this which gives the problem its one-dimensional nature. However, we do have to investigate some of the details of what happens internal to the rod when it bends.

> Actually, we do not use the continuity or momentum equation at all in our analysis of the bending of a beam! All we need is Hooke's law relating the pattern of internal stress to that of the internal strain. The rest of the argument is independent of the other ideas developed in previous sections. Instead, we introduce a little of the physics of a "shear stress" which is vital to continuum mechanics in two- and three-dimensions.   ASIDE

To describe how a beam, as shown in Figure 4.3, bends we assume a number of things: the beam is in equilibrium (that the movement has settled down) and so the internal stresses exactly balance the external forces; the beam is a Hookean solid obeying Hooke's law (4.1); the unknown deformation $y = Y(x)$ is so small from the reference state $y = 0$ that the beam is almost flat; the beam is of uniform rectangular cross-section; and no horizontal force acts on the beam.   The results are no different for other cross-sections

The stress generated by bending

Consider a small slice of a bent beam, as shown in Figure 4.4. Locally the bent beam is nearly circular, and so a line element in the unstretched

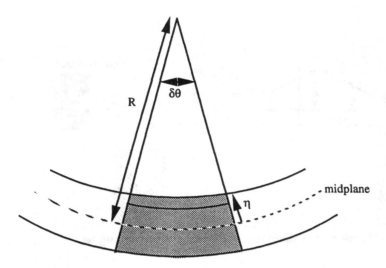

Figure 4.4: a small part of the bent beam showing the unstretched midplane.

midplane of the beam has length $R\delta\theta$ where R is the local radius of curvature of the beam. Thus the original length of all horizontal line elements in this slice of the beam was $R\delta\theta$. However, at a distance η from the midplane a line element now has the length $(R - \eta)\delta\theta$, to be compared to its unstretched length of $R\delta\theta$. Thus the local strain and hence the local stress is

$$e = \frac{(R - \eta)\delta\theta - R\delta\theta}{R\delta\theta} = -\frac{\eta}{R} \quad \Rightarrow \quad \sigma = Ee = -\frac{E\eta}{R} \ .$$

That is, there is a linear variation in the stress across the beam.

Now, integrating over the rectangular cross-section of the beam gives

$$\begin{aligned}
\text{net horizontal force} \ &= \ b \int_{-h/2}^{h/2} \sigma \, d\eta \\
&= \ b \int_{-h/2}^{h/2} -\frac{E\eta}{R} \, d\eta \\
&= \ -\frac{bE}{R} \int_{-h/2}^{h/2} \eta \, d\eta \\
&= \ 0 \ .
\end{aligned}$$

This just confirms that the midplane is indeed unstretched. If the net horizontal force were not zero then the beam would locally move to one side or another until the net horizontal forces balance.

But the pattern of internal stress does produce a twisting force, a **moment** (or **torque**) on the material. Consider the internal stresses at a cross-section X_1, as shown in Figure 4.5, which are the forces exerted

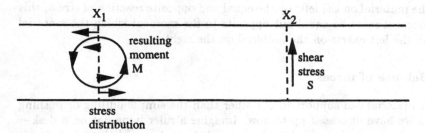

Figure 4.5: the internal stress in a bent beam generate a twisting force; this is related to the shear stress acting across any cross-section.

by the material immediately to the right of X_1 on the material which is immediately to the left of X_1. On the material to the left of X_1 it thus exerts a moment (relative to the midplane) of

Positive moments are anti-clockwise

$$
\begin{aligned}
M &= b \int_{-h/2}^{h/2} (-\sigma)\eta \, d\eta \\
&= b \int_{-h/2}^{h/2} \frac{E\eta}{R} \eta \, d\eta \\
&= \frac{Eb}{R} \int_{-h/2}^{h/2} \eta^2 \, d\eta \\
&= \frac{Ebh^3}{12R} \\
&= EI\frac{1}{R}
\end{aligned}
$$

where I is the **area moment of inertia**, and is given by

$$
I = \frac{bh^3}{12} = \frac{1}{12}(bh)^2\frac{h}{b} = \frac{1}{12}(\text{area})^2(\text{aspect-ratio})
$$

Lastly, the radius of curvature R of the beam is approximated by

$$
\frac{1}{R} = \frac{d^2Y/dx^2}{[1 + (dY/dx)^2]^{3/2}} \approx \frac{d^2Y}{dx^2} \, ,
$$

as the beam is nearly flat and so $\frac{dY}{dx}$ is negligible when compared to 1. Thus the bending moment is

$$M(x) = EI\frac{d^2Y}{dx^2} \ . \tag{4.5}$$

This is the bending moment which the material on the right exerts on the material on the left; by the equal and opposite reaction of stress, this moment must be equal but opposite to the moment which the material on the left exerts on the material on the right.

Balance of forces

A material can support forces other than the simple pulling or pushing as we have discussed up to now. Imagine a ruler lying flat on a desk— we can move the ruler by pressing lightly on top and then dragging it sideways. This movement in the direction along the ruler came from a force that was *applied* from above and is an example of a shear stress. To investigate the bending of a beam we define the **shear stress** S, as shown at the cross-section X_2 in Figure 4.5, to be the upwards force the material immediately to the right of a cross-section exerts on the material to the left. By Newton's third law this is, of course, equal but opposite to the shear stress which the material to the left exerts on the material to the right.

To derive the governing equations consider the balance of forces in a slice of the beam $[a, b]$ as shown in Figure 4.6. Note that the sideways

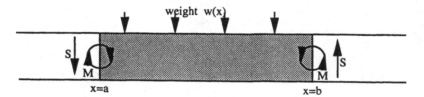

Figure 4.6: the forces acting upon a slice of a bending beam

force w acting on the beam is usually a weight and so we take positive w to be acting downwards, which is opposite to the normal sign convention.

Now if the beam is in equilibrium then all the vertical forces must exactly balance. Thus

$$S(b) - S(a) - \int_a^b w \, dx = 0$$

$$\Rightarrow\ [S]^b_{x=a} - \int_a^b w\,dx = 0$$

$$\Rightarrow\ \int_a^b \left\{ \frac{dS}{dx} - w \right\} dx = 0 \ ,$$

but this holds for all $a < b$ and so, by the slicing theorem in Section 2.1.1, the integrand must be identically zero. That is,

$$\frac{dS}{dx} = w \ . \qquad (4.6)$$

Imagine holding a ruler horizontal by one end. All this relation says (via integration) is that as x increases away from the free end, the shear stress S supports all the weight from the free end to the position x.

Twisting forces on the slice of the beam must also be in balance. Consider Figure 4.6 once again and all the moments acting about the middle of the left end of the slice. If they are to all balance then

$$-M(a) + M(b) - \int_a^b (x-a)w\,dx + (b-a)S(b) = 0$$

$$\Rightarrow\ [M]^b_{x=a} - \int_a^b (x-a)\frac{dS}{dx}\,dx + (b-a)S(b) = 0 \quad \text{by equation (4.6)}$$

upon integrating by parts

$$\Rightarrow\ \int_a^b \frac{dM}{dx}\,dx - [(x-a)S]^b_{x=a} + \int_a^b S\,dx + (b-a)S(b) = 0$$

$$\Rightarrow\ \int_a^b \left\{ \frac{dM}{dx} + S \right\} dx = 0 \ .$$

Once again, this is true for all $a < b$ and so

$$\frac{dM}{dx} = -S \ , \qquad (4.7)$$

for all x in the beam.

The beam equation

Now combine the three governing equations (4.5–4.7) into one equation for the deflection $Y(x)$. Thus

$$EI\frac{d^2Y}{dx^2} = M$$

$$\Rightarrow \quad EI\frac{d^3Y}{dx^3} = \frac{dM}{dx} = -S$$

$$\Rightarrow \quad EI\frac{d^4Y}{dx^4} = -\frac{dS}{dx} = -w \ ,$$

and so derive the **beam equation**

$$\frac{d^4Y}{dx^4} = -\frac{w(x)}{EI} \ , \tag{4.8}$$

a fourth-order differential equation for the deflection $Y(x)$.

ASIDE   This derivation assumes that EI =constant. If this is not the case, for example, if the beam thickness or shape varies, then the more general governing differential equation is $\frac{d^2}{dx^2}\left(EI\frac{d^2Y}{dx^2}\right) = -w(x)$.

But what about the boundary conditions which are necessary to calculate a unique solution? Because the beam equation is fourth-order we need four boundary conditions: two at each end of the beam. The boundary conditions depend upon how the beam is supported—the three main conditions on the end of a beam are shown in Figure 4.7.

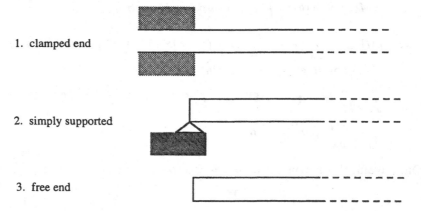

Figure 4.7: The three main sorts of conditions which apply at the ends of a beam

1. *Clamped end*: in this case the boundary conditions are simply that

$$Y = \frac{dY}{dx} = 0$$

which just says that the clamp fixes the end of the beam and its slope.

2. *Simply supported*: in this case $Y = 0$ by the support, and also $M = 0$ since a non-zero moment would cause the beam to pivot about the support; thus

$$Y = \frac{d^2Y}{dx^2} = 0 \ .$$

3. *Free end*: in this case $M = 0$ and $S = 0$ as there is no material on the other side of the end of the beam on which to exert a force; thus

$$\frac{d^2Y}{dx^2} = \frac{d^3Y}{dx^3} = 0 \ .$$

Example 4.2 Consider a log bridge over a bottomless chasm with cliffs on the sides at $x = \pm 1$. Assume it is of constant thickness, that is $I = $ const, how is it bent under its own weight? that is when $w(x) = $ const. In this case the beam equation may be integrated (four times) easily to give the general solution

$$Y = \frac{-w}{24EI}x^4 + Ax^3 + Bx^2 + Cx + D \ .$$

The simply supported boundary conditions, that $Y = \frac{d^2Y}{dx^2} = 0$ at $x = \pm 1$, determine the four constants A, B, C and D so that the deflection of the log is

$$Y = \frac{-w}{24EI} \left(x^4 - 6x^2 + 5 \right) \ .$$

Observe that for this deflection

$$M = EI\frac{d^2Y}{dx^2} = \frac{w}{2} \left(1 - x^2 \right) \quad \text{and} \quad S = -\frac{dM}{dx} = wx$$

so that the internal moment M is largest in the middle of the log. Since M measures the pattern of internal stresses, compressive on top of the log and extensive on the bottom of the log, we see that if the log is to break then it will tend do so where M is largest, namely near $x = 0$ which is the middle of the log.

ASIDE

Observe that the deflection of the log, or in general a beam, is proportional to $1/I$; thus a large moment of inertia I is desirable if beam deflections are to be minimised. For a rectangular beam we derived $I = \frac{1}{12}(\text{area})^2(\text{aspect-ratio})$, and thus for a given amount of material (area) we can make I large by increasing the aspect-ratio, h/b. Thus the strongest beams for a given weight are thin and high, which is one of the principles lying behind the use of I-beams in constructions.

Exercises

EASY

Problem 4.4 Consider a handy-person's tape measure extended horizontally to some length L: it bends gently down if it is not extended too far, but if it is extended far enough it will suddenly snap downwards, bending sharply near where it is held. This is an example of a bending beam, albeit of a curious cross-sectional shape. Assume the tape measure bends under its own uniform weight, with a clamped end at $x = 0$ and a free end at $x = L$. Find the deflection of such a beam. Calculate the resulting moment M and hence argue that when the tape measure fails to support its own weight near the horizontal, it will first fail near where it is held.

MEDIUM

Problem 4.5 Consider a beam which is *not* in equilibrium, but which is vibrating in some fashion—perhaps due to a time-dependent applied weight $w(x,t)$. The relation between bending and internal moment, equation (4.5), must still hold.

(a) For a beam of mass density ρ and vertical acceleration $\frac{\partial^2 Y}{\partial t^2}$ derive appropriately modified versions of equations (4.6–4.7). Use the principles that any net vertical force must correspond to a net vertical acceleration, and that any net moment about the left end must correspond to a net angular acceleration about the left end.

(b) Hence derive the **beam equation** governing time-dependent deflections under load:

$$\frac{\partial^2 Y}{\partial t^2} = -EI\frac{\partial^4 Y}{\partial x^4} - w(x,t) \ .$$

(c) Ignoring any load w, find the modes of vibration of a beam which is simply supported at $x = 0$ and $x = L$. *Hint*: substitute $Y(x,t) = \cos(\omega t)y(x)$ and find that solutions only exist for frequencies $\omega_n = (n\pi/L)^2\sqrt{EI/\rho}$.

(d) HARDER Find the frequencies of vibration of a beam which is clamped at $x = 0$ and has a free end at $x = L$. *Hint*: write the frequencies in terms of the nth solution ζ_n to $\cos\zeta = -\text{sech}\,\zeta$, for which we find $\zeta_1 = 1.8751 = 0.5969\pi$, $\zeta_2 = 4.6941 = 1.4942\pi$, $\zeta_3 = 7.8548 = 2.5002\pi$, *etc*. Using a tape measure, confirm experimentally that the frequency of the fundamental mode is indeed proportional to $1/L^2$. Estimate EI/ρ for your tape measure (mine is about $1.5m^4/s^2$).

Note that the above beam equation for sideways vibrations is quite different to the wave equation for the sideways vibration of a string under tension or for the longitudinal waves in a rod. This is most clearly seen in the frequency dependence upon length: $\omega \propto 1/L$ for the wave equation and $\omega \propto 1/L^2$ for the beam equation. Thus sideways waves on a beam travel much differently than do longitudinal waves.

4.3   Constitutive relations: visco-elasticity

The stress-strain relationships used up to now, namely Hooke's law (4.1) and the ideal gas equation of state (3.3), are just the simplest of ideal-isations. Fortunately, they are sufficiently accurate for many practical purposes. However, the behaviour of many materials is significantly different; indeed, the range of possibilities for the stress-strain relationship is horrendous.

For a **solid**, the stress-strain relation is generally: nonlinear, $\sigma = f(e)$ as drawn earlier in Figure 4.1; depends upon the rate-of-strain, $\sigma = f(e, \frac{\partial e}{\partial t})$; or may depend upon the previous history, $\sigma = \int_0^t f\left[e(x, \tau)\right] d\tau$.

This dependence upon the rate-of-strain is particularly true for biological materials like muscles or ligaments. There a slow extension produces less stress than a fast extension. It is for this reason that muscle stretching exercises should be done slowly; so-called "bouncing" stretches must be avoided if one is to avoid tearing a muscle or ligament.

A **fluid** is generally characterised by a complete lack of "memory", that is $\sigma = -p + f\left(\frac{\partial e}{\partial t}\right)$ where the dependence upon only $\frac{\partial e}{\partial t}$ shows that the stress is independent of the net strain e and depends only upon its rate-of change. For example, the simplest non-ideal fluid is a **Newtonian fluid** (e.g. air or water) for which

$$\sigma = -p + \eta \frac{De}{Dt} \approx -p + \eta \frac{\partial v}{\partial x}$$

where η is some constant of the fluid. The term $\eta \frac{\partial v}{\partial x}$ appears because in a real fluid an expansion ($\frac{\partial v}{\partial x} > 0$) or a contraction ($\frac{\partial v}{\partial x} < 0$) is nearly but not quite reversible; and this term represents the dissipative processes at work. Just like Hooke's law for a solid, this relation is just a tractable starting point for the investigation of the behaviour of real fluids. In general, their stress-strain relation is nonlinear and also depends upon the previous history of the fluid.

A further complication is that many materials have a mixed be-haviour, they may behave either like solids or like liquids depending upon the time-scale of the deformation. For example, **silly putty** bounces if thrown to the floor, but oozes flat if left sitting on a desk: on a time-scale less than a second it behaves like an elastic solid; while on a time-scale greater than about a minute it behaves like a fluid! For another example, **glass** is certainly solid on the time-scale of our lives, but over a time-scale of a thousand years it behaves like a liquid—the glass in many old church windows is very gradually flowing down the window pane!

To quantify some of these ideas we investigate next the consequences of some microscopic models of the behaviour of non-Hookean solids. These are models of **visco-elasticity**. These particular models still exhibit a linear relation between the stress and the strain; however, they show how dependence upon the previous history may arise. Further information may be found in the book by Fung [4, Chapts. 1 and 15].

4.3.1 Maxwell's model

We construct models of visco-elastic behaviour from the two microscopic components illustrated in Figure 4.8: a Hookean spring with a stress-

Hookean spring   dashpot

Figure 4.8: The symbols of the two microscopic components of a visco-elastic model: a Hookean spring; and a Newtonian dashpot.

strain relation of $\sigma = Ee$; and a Newtonian dashpot with a stress-strain relation of $\sigma = \eta\dot{e}$, where η is the coefficient of viscosity. Note that in this section we frequently use the overdot to denote differentiation with respect to time. The spring represents interatomic elasticity in the material under consideration. The dashpot—think of a pot of honey or treacle in which the plunger slowly moves—represents a fluid-like creep due to the rearrangement of atoms or the movement of crystal dislocations within the material. Of course, we imagine that a purely elastic bar is made up of millions of these tiny components connected serially, as shown in Figure 4.9. Similarly, we imagine that an ideal Newtonian "fluid", albeit one-dimensional, is made up of millions of dashpots connected serially, as is also shown in the figure. However, most materials are not precisely either of these ideals.

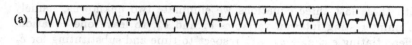

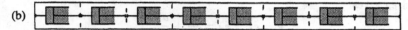

Figure 4.9: (a) Many microscopic Hookean springs connected together to make a Hookean elastic bar; (b) many Newtonian dashpots connected together to make a Newtonian "fluid."

Instead, most materials, like silly putty, share a mixture of these properties. **Maxwell's model** is to assume that the fundamental atomic unit of the material is a spring and a dashpot connected in series; millions of these units are then imagined to be attached end-to-end to form a bar of the material as is shown in Figure 4.10. Our aim here is to relate the overall strain, or extension e, of each unit to the overall stress, σ, to which it is subjected—this being the visco-elastic stress-strain relation.

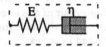

Figure 4.10: Maxwell's model of visco-elasticity is to connect a dashpot and spring in series as the basic unit of the material.

Now the total extension of each unit is clearly the sum of the extension of its two components: the spring's extension e_s, and the dashpot's extension e_d. Thus

$$e = e_s + e_d \, .$$

A little more subtle is the fact that the stresses to which the two components are subjected are the same, and furthermore, they are the same as the stress, σ, acting on the unit as a whole. This is because the stress, being an internal force, can only be transmitted from one end of the unit to the other via each of the two components in sequence. Hence the spring and the dashpot evolve according to their respective stress-strain relationships, namely

$$\sigma = E e_s \quad \text{and} \quad \sigma = \eta \dot{e}_d \, .$$

With these last three equations we eliminate two unknowns, namely the uninteresting internal variables e_s and e_d, to leave one equation. Differentiating $e = e_s + e_d$ with respect to time and substituting for $\dot{e}_s$ and $\dot{e}_d$ gives

$$\dot{e} = \dot{e}_s = \dot{e}_d = \frac{\dot{\sigma}}{E} + \frac{\sigma}{\eta} \ .$$

This is normally rearranged to the form

$$\sigma + \frac{\eta}{E}\dot{\sigma} = \eta\dot{e} \ , \tag{4.9}$$

a differential stress-strain relationship.

ASIDE   The arguments just employed should sound familiar to anyone who has studied electronics. There is an exact analogy between electrical circuits, with resistor and capacitor components and the resulting behaviour of current and voltage, and these microscopic models of visco-elasticity. The analogy is as follows: strain on a component is analogous to voltage drop; stress is analogous to the electric current; an elastic spring with spring constant E is analogous to a resistor with conductivity E (resistance $1/E$); and a dashpot with viscosity η is analogous to a capacitor with capacitance η. Then the behaviour of the above microscopic unit of a spring and a dashpot in series is directly analogous to the behaviour of a resistor and a capacitor connected in series.

Creep test

Under the application of a given applied force, stress σ, a bar of visco-elastic material will deform in some manner, as measured by the strain e. The **creep test** is the idealisation where instantaneously, at time $t = 0$, a fixed stress is applied and thereafter maintained on the bar of material; the resulting deformation response $e(t)$, the **creep**, is characteristic of the visco-elasticity of the material. The general response, when starting from an unstrained state $e(0) = 0$, is obtained simply by integrating equation (4.8) to give

$$\eta e(t) = \int_0^t \sigma(\tau)\,d\tau + \frac{\eta}{E}\sigma(t) \ .$$

The simplest creep test of suddenly applying a fixed constant stress,

$$\sigma(t) = \begin{cases} 0 & \text{for } t \le 0 \\ \sigma_0 & \text{for } t > 0 \end{cases}$$

gives rise to the response

$$e(t) = \sigma_0 c(t) \quad \text{where} \quad c(t) = \begin{cases} 0 & \text{for } t < 0 \\ \left(\frac{t}{\eta} + \frac{1}{E}\right) & \text{for } t > 0 \end{cases}$$

as shown in Figure 4.11. The response has two components: first there

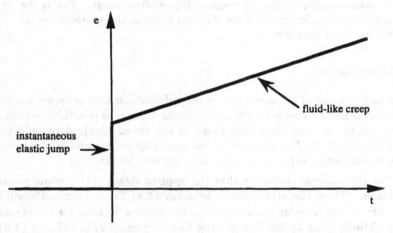

Figure 4.11: the deformation, $e(t)$, of a bar of Maxwell's material when subject to a suddenly applied stress σ_0.

is the instantaneous elastic deformation of the Hookean spring; which is then followed by the "fluidic creep" of the dashpot under the continued load. In a solid, this "fluid-like" creep could arise from the movements of the atomic dislocations (the imperfections) in the crystal structure. The function $c(t)$ is called the **creep function** and may be used to determine the deformation caused in response to any applied history of stress $\sigma(t)$, namely

$$e(t) = \int_0^t c(\tau) \dot{\sigma}(t - \tau) \, d\tau \ . \tag{4.10}$$

This model of a visco-elastic material will clearly break down under a continuing load as the resulting deformation will continue to grow for large time and so this linear theory of visco-elasticity (based on small displacements) must eventually become invalid.

In the electrical circuit analogue, the above creep test is equivalent to   ASIDE
forcing a fixed current (the fixed stress) through a resistor and a capacitor

(the spring and the dashpot respectively) and measuring the voltage drop (the strain) across the combination. Clearly, the voltage drop across the resistor will be constant, $V = IR$, while that across the capacitor will increase linearly in time as the charge accumulates, $dV/dt = I/C$, due to the fixed current. Another characteristic of the coupled resistor-capacitor is the current which flows through them when a fixed voltage is applied across the pair. In visco-elasticity the analogue is to apply a fixed strain, or deformation, and then to measure the resulting stress. This is the following relaxation test and the resultant stress is also characteristic of the nature of the material.

Relaxation test

The complement of the creep test is the **relaxation test** where a bar of material is instantaneously deformed (i.e. as quickly as possible) to some fixed length, e_0, and then held there as the stress required to hold it, $\sigma(t)$, is measured. The required stress generally decays and it measures how the material adjusts, or relaxes, to its new length.

For the moment, consider that the applied strain $e(t)$ is some given function. Then the stress-strain relation (4.9) for a Maxwell's solid appears as a first-order linear differential equation in time for the stress $\sigma(t)$. Multiplying by the integrating factor $\frac{E}{\eta} \exp(Et/\eta)$ reduces (4.9) to

$$\frac{d}{dt}\left[\exp(Et/\eta)\sigma\right] = E\exp(Et/\eta)\dot{e} \ .$$

Integrating this with respect to time from an unstressed state at time $t = 0$ to the current time t gives

$$\sigma = E\exp(-Et/\eta)\int_0^t \exp(E\tau'/\eta)\dot{e}(\tau')\,d\tau'$$

which, for ease of interpretation, becomes

$$\sigma = E\int_0^t \exp(-E\tau/\eta)\dot{e}(t-\tau)\,d\tau$$

upon changing the integration variable to $\tau = t - \tau'$. In this form observe that, since t is the present time, τ measures time before the present in that $\dot{e}(t-\tau)$ refers to the rate-of-strain a time τ ago. Thus the integral describes a dependence upon the previous history of the strain with an exponentially decaying weight of $\exp(-E\tau/\eta)$. Assuming that $e^{-4} = 0.0183$ is negligible compared to 1 we see that this memory of the earlier state of strain lasts a time of $t \approx 4\eta/E$.

Integrating this memory integral by parts, assuming that the initial strain is $e(0) = 0$, gives

$$\sigma = Ee(t) - \frac{E^2}{\eta} \int_0^t \exp(-E\tau/\eta) e(t - \tau) \, d\tau$$

which shows that the stress is the Hookean elastic stress, $Ee(t)$, less an adjustment due to the creep of the dashpot. Applying the simplest strain of the sudden fixed displacement

$$e(t) = \begin{cases} 0 & \text{for } t \leq 0 \\ e_0 & \text{for } t > 0 \end{cases}$$

gives rise to the following stress in the material

$$\sigma(t) = e_0 k(t) \qquad \text{where} \quad k(t) = \begin{cases} 0 & \text{for } t < 0 \\ E \exp(-Et/\eta) & \text{for } t > 0 \end{cases}$$

as shown in Figure 4.12. Once again the response has two components: there is first the instantaneous elastic deformation of the spring to accommodate the applied deformation; this is followed by an exponential decay, a relaxation, in the stress as the dashpot stretches to the extent of the deformation. The function $k(t)$ is called the **relaxation function** and may be used to determine the stress caused by any applied history of strain $e(t)$, namely

$$\sigma(t) = \int_0^t k(\tau) \dot{e}(t - \tau) \, d\tau \qquad\qquad (4.11)$$

as seen above.

Wave propagation

Another characteristic of the model is how the visco-elasticity affects the propagation of waves along a bar of the material. For simplicity, we only consider the propagation of small amplitude waves governed by the linearised equations. Since waves involve spatial variations as well as the time variations of the previously uniform straining, we return to using the partial derivative notation rather than the overdots.

Maxwell's stress-strain relation (4.9), $\sigma + \frac{\eta}{E} \frac{\partial \sigma}{\partial t} = \eta \frac{\partial e}{\partial t}$, is differentiated with respect to x to give

$$\frac{\partial \sigma}{\partial x} + \frac{\eta}{E} \frac{\partial^2 \sigma}{\partial x \partial t} = \eta \frac{\partial^2 e}{\partial x \partial t} \ .$$

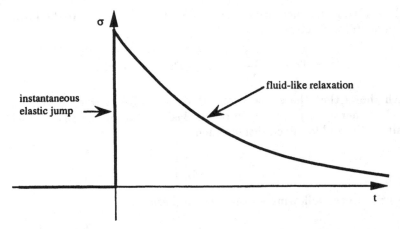

Figure 4.12: the relaxation of the stress, $\sigma(t)$, in a bar of Maxwell's material when subject to a suddenly applied deformation e_0.

The linearised momentum equation $\rho\frac{\partial v}{\partial t} = \frac{\partial \sigma}{\partial x}$, from equation (3.2), is then used to eliminate the spatial derivatives of the stress σ in favour of time derivatives of the velocity, thus

$$\rho\frac{\partial v}{\partial t} + \frac{\rho\eta}{E}\frac{\partial^2 v}{\partial t^2} = \eta\frac{\partial^2 e}{\partial x \partial t} \ .$$

Cauchy's infinitesimal strain $e = \frac{\partial u}{\partial x}$, equation (4.2), is differentiated with respect to time to give $\frac{\partial e}{\partial t} = \frac{\partial^2 u}{\partial x \partial t} = \frac{\partial v}{\partial x}$ by the linearised relation (4.3) between the displacement field u and the velocity field v. Eliminating e from the above equation and dividing through by the coefficient of $\frac{\partial^2 v}{\partial t^2}$ then gives

$$\frac{\partial^2 v}{\partial t^2} + \frac{E}{\eta}\frac{\partial v}{\partial t} = \frac{E}{\rho}\frac{\partial^2 v}{\partial x^2} \ .$$

This is basically a wave equation, with the elastic wave speed $c = \sqrt{E/\rho}$, but modified by the, as it turns out, dissipative term $\frac{E}{\eta}\frac{\partial v}{\partial t}$.

The above equation has constant coefficients and so we expect a solution may be found in terms of exponentials, possibly complex. This is so, but to put the solution in the form of a travelling wave we pose at the outset that a solution may be found in the form

$$v = Ae^{-\nu t}\sin(kx - \omega t)$$

where the $\sin(kx - \omega t)$ factor describes a sinusoidal wave of wavelength $L = 2\pi/k$ and propagating with speed ω/k. The $e^{-\nu t}$ factor gives an exponential decay to the wave's amplitude due to the dissipative processes of the microscopic mechanics. Substituting this into the above equation gives

$$-\omega^2 A e^{-\nu t} \sin(kx - \omega t) + 2\omega\nu A e^{-\nu t} \cos(kx - \omega t) +$$
$$+\nu^2 A e^{-\nu t} \sin(kx - \omega t) - \frac{E}{\eta}\left(\omega A e^{-\nu t} \cos(kx - \omega t) - \right.$$
$$\left. -\nu A e^{-\nu t} \sin(kx - \omega t)\right) = -k^2 \frac{E}{\rho} A e^{-\nu t} \sin(kx - \omega t)$$

which upon grouping like terms is arranged as

$$\left[-\omega^2 + \nu^2 - \frac{\nu E}{\eta} + \frac{E k^2}{\rho}\right] A e^{-\nu t} \sin(kx - \omega t) +$$
$$+\omega\left[2\nu - \frac{E}{\eta}\right] A e^{-\nu t} \cos(kx - \omega t) = 0 \ .$$

Setting the coefficients of this equation to zero gives two equations: one to determine the rate of exponential decay

$$\nu = \frac{E}{2\eta} \ ;$$

and the other the frequency as a function of wavenumber k,

$$\omega = \sqrt{\frac{E}{\rho}k^2 - \left(\frac{E}{2\eta}\right)^2} \ .$$

Thus Maxwell's model of visco-elasticity possesses waves that decay like $\exp(-Et/2\eta)$ and propagate at a wave speed which is slightly depressed,

$$c = \frac{\omega}{k} = \sqrt{\frac{E}{\rho} - \left(\frac{E}{2\eta k}\right)^2} \ ,$$

from that of purely elastic waves. This wave speed correction is most marked for long waves, small $k = 2\pi/L$; indeed for very long waves the frequency becomes complex which indicates that the interpretation of the dynamics as a long propagating wave no longer holds. Instead, for large spatial structures the dynamics are purely dissipative.

4.3.2 Kelvin's model (Voigt's model)

Kelvin's model of visco-elasticity is to assume that the fundamental atomic unit of the material is a spring and a dashpot connected in *parallel*; millions of these units are then imagined to be attached end-to-end to form a bar of the material as is shown in Figure 4.13. The aim here is to relate the overall strain, or extension e, of each unit to the stress, σ, to which it is subjected—this being the visco-elastic stress-strain relationship.

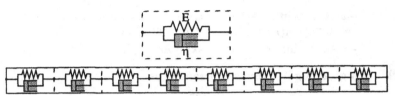

Figure 4.13: Kelvin's model of visco-elasticity is to connect a dashpot and spring in parallel as the basic unit of the material.

The total extension of each unit is clearly the same as the extension of its two components: the spring, and the dashpot. However, the stress experienced across each unit must be apportioned between the two: the spring's stress σ_s, and the dashpot's stress σ_d. Thus

$$\sigma = \sigma_s + \sigma_d \ .$$

Now, the stresses in the spring and the dashpot must obey their respective stress-strain relationships, namely

$$\sigma_s = Ee \qquad \text{and} \qquad \sigma_d = \eta\dot{e} \ .$$

With these last three equations we substitute for the two uninteresting unknowns, namely the internal variables σ_s and σ_d, to derive the following equation

$$\sigma = Ee + \eta\dot{e} \ , \tag{4.12}$$

relating the extension e to the total stress σ of a unit.

ASIDE

In the analogy between electrical circuits and these microscopic models of visco-elasticity the behaviour of the above microscopic unit of a spring and a dashpot in parallel is directly analogous to the behaviour of a resistor (resistance "$1/E$") and a capacitor (capacitance "η") connected in parallel.

Creep test

The **creep test** is to instantaneously, at time $t = 0$, apply a fixed stress on a bar of the material which initially is unstrained, $e(0) = 0$. That is

apply
$$\sigma(t) = \begin{cases} 0 & \text{for } t \leq 0 \\ \sigma_0 & \text{for } t > 0 \end{cases}$$

The resulting deformation response $e(t)$, the **creep**, is characteristic of the visco-elastic behaviour of the material. The general response is obtained by solving the constant coefficient differential equation (4.12) with $\sigma = \sigma_0$. The general solution is

$$e(t) = \frac{\sigma_0}{E} + c_1 e^{-Et/\eta} \, ,$$

which with the initial condition that $e(0) = 0$ then gives rise to the response

$$e(t) = \sigma_0 c(t) \quad \text{where} \quad c(t) = \begin{cases} 0 & \text{for } t < 0 \\ \frac{1}{E} \left(1 - e^{-Et/\eta}\right) & \text{for } t > 0 \end{cases}$$

as shown in Figure 4.14. The response shows that the initial fluid-like stretching is arrested on a time-scale $t \approx \eta/E$ by the spring in parallel taking up the stress. Once again the **creep function** $c(t)$ may be used to determine the deformation caused in response to any applied history of stress $\sigma(t)$, namely through the formula (4.10).

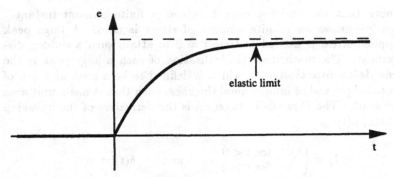

Figure 4.14: the deformation, $e(t)$, of a bar of Kelvin's material when subject to a suddenly applied stress σ_0.

ASIDE

In the electrical circuit analogue, the creep test is equivalent to forcing a fixed current (the fixed stress) through the unit and measuring the voltage drop (the strain) across the combination. Initially, all of the current will flow through the capacitor. However, as the voltage drop across the capacitor increases, $dV/dt = I/C$, more of the current will flow through the resistor. The voltage drop will continue to increase until all the supplied current passes through the resistor, $V = IR$.

Conversely, the visco-elastic relaxation test is analogous to applying a fixed voltage across the unit and measuring the resultant current (analogous to the stress). In the electric circuit, an almost instantaneous, extremely large pulse of current then would flow through the capacitor as it short-circuits the current. This pulse lasts to build up the voltage across the capacitor to that which is applied, thereafter a constant current flows through the resistor. These features are seen in Kelvin's model under an instantaneous deformation.

Relaxation test

In the relaxation test a bar of material is instantaneously deformed (i.e. as quickly as possible) to some fixed length, e_0, and then held there as the stress, $\sigma(t)$, required to hold it is measured. For a Kelvin solid the computation of the necessary stress is nearly trivial as it is given explicitly in terms of $e(t)$ by equation (4.12).

However, in trying to obtain the sudden fixed displacement

$$e(t) = \left\{ \begin{array}{ll} 0 & \text{for } t \le 0 \\ e_0 & \text{for } t > 0 \end{array} \right.$$

we note that the dashpot cannot extend a finite amount instantaneously—unless an infinite amount of stress is used. A huge peak of applied stress is needed at time $t = 0$ to attain such a sudden displacement. The mathematical idealisation of such a huge peak is the **Dirac delta function**, $\delta(t)$, which is defined to be a peak at $t = 0$ of infinite height and of infinitesimal thickness such that it has a unit area underneath. The Dirac delta function is the derivative of the unit step function, $u(t)$:

$$u(t) = \left\{ \begin{array}{ll} 0 & \text{for } t < 0 \\ 1 & \text{for } t > 0 \end{array} \right. \qquad \text{and} \qquad \delta(t) = \dot{u}(t) \ .$$

Thus the above applied stress $e = e_0 u(t)$ gives rise to the following stress in the material

$$\sigma(t) = e_0 k(t) \qquad \text{where} \quad k(t) = \left\{ \begin{array}{ll} 0 & \text{for } t < 0 \\ \eta \delta(t) + E & \text{for } t \ge 0 \end{array} \right.$$

as shown in Figure 4.15. Once again the function $k(t)$ is called the relaxation function and may be used to determine the stress caused by any applied history of strain $e(t)$, namely through the formula (4.11).

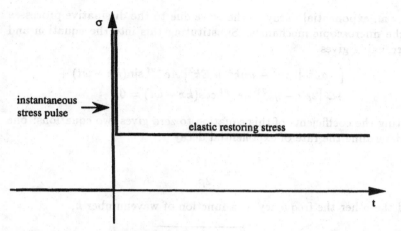

Figure 4.15: the relaxation of the stress, $\sigma(t)$, in a bar of Kelvin's material when subject to a suddenly applied deformation e_0.

Wave propagation

The propagation of waves along a bar of the material is also characteristic of the visco-elasticity. Starting with Kelvin's stress-strain relation, equation (4.12), derive

$$\frac{\partial \sigma}{\partial t} = E\frac{\partial e}{\partial t} + \eta\frac{\partial^2 e}{\partial t^2} \quad \text{upon taking } \tfrac{\partial}{\partial t}$$

$$= E\frac{\partial v}{\partial x} + \eta\frac{\partial^2 v}{\partial x \partial t} \quad \text{using } \tfrac{\partial e}{\partial t} = \tfrac{\partial^2 u}{\partial x \partial t} = \tfrac{\partial v}{\partial x}$$

$$\Rightarrow \frac{\partial^2 \sigma}{\partial x \partial t} = E\frac{\partial^2 v}{\partial x^2} + \eta\frac{\partial^3 v}{\partial x^2 \partial t} \quad \text{upon taking } \tfrac{\partial}{\partial x}$$

$$\Rightarrow \rho\frac{\partial^2 v}{\partial t^2} = E\frac{\partial^2 v}{\partial x^2} + \eta\frac{\partial^3 v}{\partial x^2 \partial t} \quad \text{via linearised momentum } \rho\tfrac{\partial v}{\partial t} = \tfrac{\partial \sigma}{\partial x}.$$

This last equation is essentially a wave equation for the velocity field v, with the elastic wave speed $c = \sqrt{E/\rho}$, but modified by the dissipative term $\eta\frac{\partial^3 v}{\partial x^2 \partial t}$.

As with Maxwell's model, to put the solutions of the equation in the form of a travelling wave we postulate that a solution may be found in the form

$$v = Ae^{-\nu t}\sin(kx - \omega t)$$

where the $\sin(kx - \omega t)$ factor describes a sinusoidal wave of wavelength $L = 2\pi/k$ and propagating with wave speed ω/k. The $e^{-\nu t}$ factor gives an exponential decay to the wave due to the dissipative processes of the microscopic mechanics. Substituting this into the equation and rearranging gives

$$\left[-\rho\omega^2 + \rho\nu^2 - \nu\eta k^2 + Ek^2\right] Ae^{-\nu t}\sin(kx - \omega t) +$$
$$+\omega\left[2\rho\nu - \eta k^2\right] Ae^{-\nu t}\cos(kx - \omega t) = 0 \ .$$

Setting the coefficients of this equation to zero gives two equations: one to determine the rate of exponential decay

$$\nu = \frac{\eta k^2}{2\rho} \ ;$$

and the other the frequency as a function of wavenumber k,

$$\omega = \sqrt{\frac{E}{\rho}k^2 - \left(\frac{\eta k^2}{2\rho}\right)^2} \ .$$

Thus Kelvin's model of visco-elasticity possesses waves that decay like $\exp\left[-\eta k^2 t/(2\rho)\right]$—this viscous decay is very rapid for short waves (large wavenumber k) and virtually negligible for long waves (small wavenumber k). The waves propagate at a wave speed which is slightly depressed,

$$c = \frac{\omega}{k} = \sqrt{\frac{E}{\rho} - \left(\frac{\eta k}{2\rho}\right)^2} \ ,$$

from that of purely elastic waves. The wave speed correction is most marked for short waves; indeed for waves of wavelength $L = 2\pi/k < \pi\eta/\sqrt{\rho E}$ the frequency becomes complex which indicates that a short "wave" no longer propagates—it just decays.

4.3.3 More general models

The common theme in the above two models is the existence of the integral relationships, equations (4.10) and (4.11), between the history of the stress and the strain and the current state. Although they both introduce dissipation into the response of a material to, for example, propagating waves, the models differ in many of the details as can be seen from the different creep and relaxation functions, and their differing effects on long and short wavelength propagating waves. Neither of these models is quantitatively correct—the infinite delta-function forces of the relaxation response of Kelvin's model and the complete relaxation of Maxwell's model are generally not acceptable. A more reasonable model of visco-elasticity is the **standard linear solid** whose basic unit is shown in Figure 4.16. The differential stress-strain relationship for this model is

$$E_2\sigma + \eta\dot\sigma = E_1 E_2 e + \eta(E_1 + E_2)\dot e . \qquad (4.13)$$

To an applied stress it allows an instantaneous elastic deflection followed by a slow creep, as the dashpot causes the spring E_2 to relax, which is eventually arrested by the further straining of the spring E_1. To an applied deflection it responds with a large, but finite, immediate stress which is followed by a slow relaxation as the dashpot extends to take up some of the stretching.

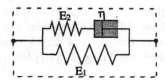

Figure 4.16: the basic unit of the standard linear solid, consisting of two springs and a dashpot.

More complex models may be built up of arbitrary numbers of springs and dashpots in an enormous variety of configurations. For example, there are five nondegenerate models which can be made from four spring and dashpot components. Such elaborate models induce a higher-order differential stress-strain relation, one which may be written in the form

$$\sigma + p_1\dot\sigma + p_2\ddot\sigma + \cdots = q_0 e + q_1\dot e + q_2\ddot e + \cdots .$$

By varying the parameters in a given model, the coefficients p_n and q_n may be almost arbitrarily varied. Thus the model and its parameters may be chosen to fit almost *any* desired creep $c(t)$ or relaxation $k(t)$ functions which determine the visco-elastic behaviour through equations (4.10) and (4.11) respectively. For example, if a desired relaxation function is to be obtained then the Laplace transform of (4.11) shows that $\hat\sigma(s) = s\hat k(s)\hat e(s)$ where transformed quantities are indicated by the $\hat{}$'s. However, the Laplace transform of the above differential stress-strain relation gives

$$(1 + p_1 s + p_2 s^2 + \cdots)\,\hat\sigma = (q_0 + q_1 s + q_2 s^2 + \cdots)\,\hat e .$$

Upon comparing these we see that the model must be chosen so that

$$\hat{k}(s) = \frac{(q_0 + q_1 s + q_2 s^2 + \cdots)}{s\left(1 + p_1 s + p_2 s^2 + \cdots\right)} .$$

That is, approximating the Laplace transform of the desired relaxation function by a rational function determines the coefficients of the microscopic model.

The relaxation function for real materials must be determined experimentally. Applying vibrations of a fixed frequency for a wide range of frequencies determines enough about the dynamic characteristics of the material to indirectly determine the creep function, at least approximately. The Fourier transform of the creep function is directly proportional to the response of the material to an applied forcing of a given frequency; the phase difference between the material's response and the applied oscillatory forcing is a direct measure of the internal dissipation in the material (see Fung [4, §1.7] for some more details).

Exercises

EASY   **Problem 4.6**

(a) Show that if two microscopic Hookean springs, with spring constants E_1 and E_2, are connected in parallel then the combination acts exactly the same as a Hookean spring with constant $E_1 + E_2$.

(b) Show that if two microscopic Hookean springs, with spring constants E_1 and E_2, are connected in series then the combination acts exactly the same as a Hookean spring with constant E where $\frac{1}{E} = \frac{1}{E_1} + \frac{1}{E_2}$.

(c) Show that if two microscopic dashpots, with viscosities η_1 and η_2, are connected in parallel then the combination acts exactly the same as a dashpot with viscosity $\eta_1 + \eta_2$.

(d) Show that if two microscopic dashpots, with viscosities η_1 and η_2, are connected in series then the combination acts exactly the same as a dashpot with viscosity η where $\frac{1}{\eta} = \frac{1}{\eta_1} + \frac{1}{\eta_2}$.

(e) Using parts (a) and (b), argue that the spring constant of a uniform bar of material should be proportional to its cross-sectional area and inversely proportional to its length.

Problem 4.7 Derive the differential stress-strain relationship, equation (4.13), for the standard linear solid. *Hint*: use the Maxwell stress-strain relation for the spring E_2 and the dashpot in series. Derive and sketch the creep function $c(t)$ and the relaxation function $k(t)$. What is the typical relaxation time exhibited in these functions? Find the decay rate and the wave speed of propagating waves in a standard linear solid.

Problem 4.8 The standard linear solid is an example of a visco-elastic model with three microscopic components, two springs and a dashpot. Draw the other three visco-elastic models with three components that *cannot* degenerate to one of the two two-component models.

DIFFICULT

MEDIUM

Chapter 5

Quasi-one-dimensional continua

Many continua are nearly but not quite one-dimensional. For example: blood flow where arteries and veins have a varying cross-section and also curve; rivers which are similar; air-flow underneath cars; the deformation of a tapering beam. To a good approximation these situations may be analysed as if they were one-dimensional.

5.1   Equations of motion

Consider a continuum, as shown in Figure 5.1, which may be curved and which has a cross-sectional area A which may vary in both time and along the continuum. Let x measure distance along the centreline, more-or-less.

> For example, we might imagine that the x-axis was a coordinate curve in an orthogonal curvilinear coordinate system.   ASIDE

Assumption 1: *the properties of the continuum are nearly constant across a cross-section.* Thus we do not have to worry about any spatial dependence other than that which occurs along the continuum.

Assumption 2: *we may largely ignore any curvature along the continuum.* That is, for our purposes the continuum is taken to be essentially straight.

111

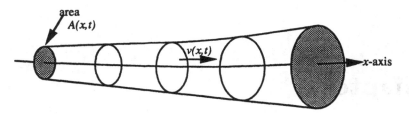

Figure 5.1: a diagram of a quasi-one-dimensional continuum, showing a varying cross-section and a curving centreline.

The material of the continuum has a density ρ *(mass/volume)* defined at each point. Note that this density is different from previous sections where ρ had dimensions *mass/length*; in this chapter we use ρ_1 to indicate such a one-dimensional density. By Assumption 1, ρ depends only upon x and not at all upon the direction perpendicular to x, thus $\rho = \rho(x,t)$. The material of the continuum also moves with some *vector* velocity v at each point. By Assumption 1, v does not vary significantly across a cross-section. Since the material cannot pass through the boundary, v must be essentially aligned along the curving x-axis. That is, $v = v(x,t)i$ where i is the local direction of the x-axis. As in previous sections, we call the scalar $v(x,t)$ the velocity of the material. Similarly for all other quantities: for example, stress $\sigma(x,t)$ is the internal force aligned along the local x-direction.

ASIDE   Sometimes the quantities may vary across the continuum, for example, the velocity in a river varies significantly across the river, $v \approx v(x,y,z,t)i$. In such a case it is usually appropriate to use the cross-sectional average as the local measure of the quantity.

5.1.1   Continuity equation

We deduce an equation which reflects the fact that material is conserved. The easiest way to proceed is to define a hypothetical, strictly one-dimensional continuum in terms of the actual three-dimensional continuum using the above assumptions. We then just write down the conservation equations for this strictly one-dimensional continuum.

Use the subscript 1 to denote quantities in the strictly one-dimensional continuum. Consider a small slice, length δx, of the continuum, the one-dimensional density is

$$\rho_1 = \frac{\text{mass}}{\text{length}}$$

$$= \frac{1}{\delta x}\,(\text{mass in } [x, x + \delta x])$$

$$\approx \frac{1}{\delta x}(\text{volume of slice})\left(\frac{\text{mass}}{\text{volume}}\right)$$

$$\approx \frac{1}{\delta x}\,(\delta x\, A)\,\rho$$

and so $\rho_1 = A\rho$. The one-dimensional density is simply the local three-dimensional density times the local cross-sectional area. As the velocity is assumed constant across a cross-section, we take $v_1 = v$. That is, the one-dimensional velocity is taken to be the local velocity in the direction of the axis of the continuum. The one-dimensional mass flux is then simply $q_1 = \rho_1 v_1 = A\rho v$. That is, the amount of material flowing through a given cross-section is proportional to not only ρ and v, as before, but also to the cross-sectional area A.

The strictly one-dimensional continuity equation (2.1), in the absence of any generation or destruction processes, asserts that

$$\frac{\partial \rho_1}{\partial t} + \frac{\partial q_1}{\partial x} = 0 \ .$$

Using the above definitions this equation becomes

$$\frac{\partial}{\partial t}\,(A\rho) + \frac{\partial}{\partial x}\,(A\rho v) = 0 \ , \tag{5.1}$$

the **quasi-one-dimensional continuity equation**.

Example 5.1 Consider an incompressible fluid ($\rho = $ constant) flowing along a pipe or channel of varying cross-sectional area. In this situation the above continuity equation reduces to $\frac{\partial A}{\partial t} + \frac{\partial}{\partial x}(Av) = 0$. Further, if the cross-sectional area is fixed, independent of t, $A = A(x)$, then $\frac{\partial A}{\partial t} = 0$ and the equation becomes $\frac{\partial}{\partial x}(Av) = 0$. Thus Av is a function of t only, say $Q(t)$ the rate at which fluid is being "pumped" along, and so

$$v = \frac{Q}{A(x)} \ .$$

Hence we recover the well known result: that the velocity of flow is highest when the cross-sectional area is smallest, and *vice-versa*.

In car traffic, when workers close off one lane of a two lane   ASIDE highway there is $A = 1$ where they are working and $A = 2$ otherwise. Thus to maintain a smooth flow of cars, instead of restricting the cars speed, from $60km/hr$ to say $30km/hr$, they should really increase it, to say $120km/hr$! However, safety considerations are paramount in practise.

5.1.2   Momentum equation

We now derive an equation which ensures that momentum only appears and disappears by being carried by the material and by the action of forces.

Just as before there are two types of forces: stress and body force. In three-dimensional mechanics a stress is taken to be a force per area, just like pressure. Thus the total stress acting through a cross-section of area A is given by $\sigma_1 = A\sigma$. The body forces will be, for the moment, just left as F_1 and defined to be the total body force applied to a cross-section.

We leave body forces denoted by F_1 because their form depends upon the nature of the external agency causing them. For example, gravity applies a body force $F = -g\rho$ to every point in the continuum which, when integrated over a cross-section, gives a one-dimensional body force of $F_1 = FA = -gA\rho$. Compare this with the friction between a fluid and the walls of a pipe which causes a retarding body force, $F = -Cv$ say, at each point of the perimeter. The length of the perimeter is proportional to $\sqrt{A}$, and therefore this gives a one-dimensional body force of $F_1 \propto \sqrt{A}F = -C\sqrt{A}v$. Observe the difficulty that the relation between F_1 and the physical force F depends upon the nature of the original force.

The strictly one-dimensional momentum equation (3.1), which reflects the conservation of momentum, is

$$\frac{\partial}{\partial t}\left(\rho_1 v_1\right) + \frac{\partial}{\partial x}\left(\rho_1 v_1^2\right) = F_1 + \frac{\partial \sigma_1}{\partial x}$$

$$\Rightarrow \quad \frac{\partial}{\partial t}(A\rho v) + \frac{\partial}{\partial x}\left(A\rho v^2\right) = F_1 + \frac{\partial}{\partial x}(A\sigma)$$

$$\Rightarrow \quad \frac{\partial(A\rho)}{\partial t}v + A\rho\frac{\partial v}{\partial t} + \frac{\partial(A\rho v)}{\partial x}v + A\rho v\frac{\partial v}{\partial x} = F_1 + \frac{\partial}{\partial x}(A\sigma)\ .$$

This last form shows how the equation may be simplified by using the continuity equation (5.1). The first and the third term on the left-hand-side of the above equation combine to give zero, and hence

$$A\rho\left[\frac{\partial v}{\partial t} + v\frac{\partial v}{\partial x}\right] = F_1 + \frac{\partial}{\partial x}(A\sigma)\ , \tag{5.2}$$

which is the **quasi-one-dimensional momentum equation** .

In fluids, which will be our main application of these ideas, if we neglect internal viscosity (friction) then the stress σ is simply the negative of the fluid pressure p. However, it is incorrect to substitute this immediately into the momentum equation (5.2); things are more subtle than this. It turns out that the appropriate momentum equation in these circumstances is

$$\frac{\partial v}{\partial t} + v\frac{\partial v}{\partial x} = \frac{F_1}{A\rho} - \frac{1}{\rho}\frac{\partial p}{\partial x} \qquad (5.3)$$

(as if $\frac{\partial}{\partial x}(Ap) = A\frac{\partial p}{\partial x}$). The reason for this, as explained in more detail below, is that, in the presence of variations in the cross-sectional area, a gradient of pressure produces an along-axis (longitudinal) force where the pressure is applied to the sloping wall of the pipe—this longitudinal force is resisted by the pipe and thus generates an equal and opposite contribution within the body force F_1 on the fluid.

An argument goes as follows. Consider a small interval of length δx over which the cross-sectional area varies from A to $A+\delta A$. This change in area is spread around the perimeter, of length P say, and so the net change in radius of the pipe is $\delta A/P$. The normal to the wall of the pipe, the direction in which pressure will act, then slopes backwards by an amount $\delta A/(P\delta x)$. Thus the fluid exerts a pressure of $-p\delta A/(P\delta x)$ on each point around the pipe, and so exerts a net force of $-p\delta A/\delta x \approx -p\frac{\partial A}{\partial x}$. By Newton's third law this force must be returned by the wall onto the fluid and so causes a contribution to the net body force so that $F_1 = F_1' + p\frac{\partial A}{\partial x}$ where F_1' includes all the other body forces. Thus

ASIDE

$$F_1 - \frac{\partial(Ap)}{\partial x} = F_1' + \frac{\partial A}{\partial x}p - \frac{\partial(Ap)}{\partial x} = F_1' - A\frac{\partial p}{\partial x}$$

as used to write down equation (5.3). Note that although the argument which we have used is only approximate, the result is exact.

5.2 Blood flow

Part of the **cardio-vascular system** is illustrated in Figure 5.2. It shows the heart connected to the main arteries, the small arteries and thence to the capillaries; and also gives typical sizes and velocities of blood flow in the various regions. We concentrate on the description of the flow in the aorta and the larger arteries.

Blood moves because of the pressure developed in the heart during pumping. Blood itself is a mildly complicated fluid (being a suspension

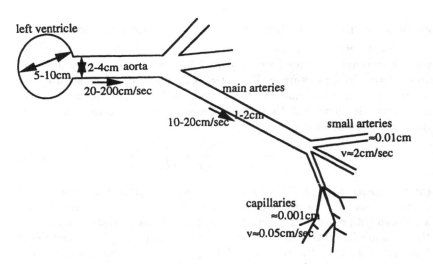

Figure 5.2: a schematic diagram of part of the cardio-vascular system.

of plate-like blood cells) but to a good approximation it is incompressible, that is ρ = constant. In the larger arteries the complicated nature of blood is not particularly significant and to a fair approximation the stress-strain relation is simply $\sigma = -p$ where p is the blood pressure.

ASIDE   This is not at all true in the small arteries and the capillaries where the presence of the plate-like blood cells is significant. In the small arteries the stress depends upon the previous history of strain, while in the capillaries, whose diameter is of the same size as the blood cells, it even becomes difficult to treat the flow as a pure fluid.

A further complication is that the arteries themselves have a rather complex structure, being made up of layers of different kinds of material, and each layer may have a directional characteristic—either along or around the artery. Basically the arteries are flexible and elastic and so interact with the blood flow, but also they contain muscle which may help to pump the blood.

5.2.1   Elastic arteries

Consider a long, cylindrical, elastic artery with cross-sectional area $A(x, t)$ which varies depending upon the blood flow, as might be shown in Figure 5.1.

Unless you
cut yourself.   Within the body, blood is conserved and so the continuity equa-

tion (5.1) holds. Here the density is constant and so

$$\frac{\partial A}{\partial t} + \frac{\partial (Av)}{\partial x} = 0 \ .$$

The arteries are essentially circular in cross-section, $A = \pi R^2$ where $R(x,t)$ is the radius of the artery, and substituting for A gives

$$2\pi R \frac{\partial R}{\partial t} + 2\pi R \frac{\partial R}{\partial x} v + \pi R^2 \frac{\partial v}{\partial x} = 0$$

$$\Rightarrow \quad \frac{\partial R}{\partial t} + v \frac{\partial R}{\partial x} + \frac{1}{2} R \frac{\partial v}{\partial x} = 0 \ . \tag{5.4}$$

Furthermore, the momentum equation (5.3) must hold. Here we will ignore body forces and take $F_1 = 0$. Gravity could be included but actually it would make no difference to the dynamics of the flow; it only changes the pressure field.

Now the artery walls are elastic: thus if the pressure increases locally then the radius R increases, that is the artery expands. Hence, there exists a relation between p and R at each point along the artery, a sort of "equation of state". To some level of approximation, postulate the linear relation

$$p = p_* + \alpha (R - R_*)$$

Hooke's law for arteries!

where p_* is the pressure which corresponds to the radius R_* and $\alpha > 0$ is a constant which depends upon the make-up of the artery—how thick it is, how elastic it is, etc. Substituting for the pressure in the momentum equation (5.3) then gives

$$\frac{\partial v}{\partial t} + v \frac{\partial v}{\partial x} = -\frac{\alpha}{\rho} \frac{\partial R}{\partial x} \ . \tag{5.5}$$

Equations (5.4–5.5) are two nonlinear, partial differential equations for the two unknowns R and v.

The standard procedure is to look now for solutions of the linearised equations. Assume that $v = \hat{v}(x,t)$ and $R = R_* + \hat{R}(x,t)$ where $\hat{v}$ and $\hat{R}$ are "small" quantities. That is, we assume the blood flow is not too fast and the artery is not deformed too much. Substituting these into the two governing equations and neglecting products of "small" terms readily gives

$$\frac{\partial \hat{R}}{\partial t} + \frac{1}{2} R_* \frac{\partial \hat{v}}{\partial x} = 0 \quad \text{and} \quad \frac{\partial \hat{v}}{\partial t} = -\frac{\alpha}{\rho} \frac{\partial \hat{R}}{\partial x} \ .$$

Starting with $\frac{\partial}{\partial t}$ of the second of these equations

$$
\begin{aligned}
\frac{\partial^2 \hat{v}}{\partial t^2} &= -\frac{\alpha}{\rho} \frac{\partial}{\partial x} \left(\frac{\partial \hat{R}}{\partial t} \right) \\
&= -\frac{\alpha}{\rho} \frac{\partial}{\partial x} \left(-\frac{1}{2} R_* \frac{\partial \hat{v}}{\partial x} \right) \quad \text{(by the first equation)} \\
&= \frac{\alpha R_*}{2\rho} \frac{\partial^2 \hat{v}}{\partial x^2} \ ,
\end{aligned}
$$

we derive the wave equation

$$
\frac{\partial^2 \hat{v}}{\partial t^2} = c_*^2 \frac{\partial^2 \hat{v}}{\partial x^2}
$$

where $c_* = \sqrt{\frac{\alpha R_*}{2\rho}}$ is a constant. D'Alembert's general solution

$$
\hat{v} = f(x - c_* t) + g(x + c_* t) \ ,
$$

describes disturbances travelling to the right and to the left, unchanging in form, with speed c_*.

Thus, when the heart pumps, producing a pulse of pressure, this pulse propagates along the arteries with speed c_*. This accounts for the just perceptible time lag between feeling your heart beat and feeling the pulse in your wrist. Note that the blood itself does not move with this velocity: blood gets caught in a pulse, is moved forward some distance, and then is left behind to wait for the next pulse.

ASIDE   One prediction from this theory is that if the pulses are unidirectional (all going in one direction with no reflection), for example $g = 0$ and $f \neq 0$, then $p \propto \hat{v} \propto \hat{R}$; which is in reasonable agreement with experiment.

ASIDE   The elasticity of an artery is important—if arteries were not elastic, then each pump of the heart would cause a virtually instantaneous rise in blood pressure throughout the body. We would then be forever juddering as each stroke of the heart stiffens up the whole body.

5.2.2 Active arteries

Arteries are not purely passive: they do have muscles, primarily rings of muscles around the artery. Muscles are a very difficult material to analyse. However, their action may be appreciated via the following simple model.

Postulate that a muscle, when it attempts to contract, simply produces an extra pressure term in the p–R relationship. That is, assume

$$p = p_* + \alpha(R - R_*) + P(x,t) \,,$$

where $P(x,t)$ is a prescribed function of distance along the artery x and in time t. This muscle action in the artery will be linked to the involuntary action of the heart.

Observe that if the pressure in the artery is held constant, $p = p_*$, then this postulate predicts $R = R_* - \frac{1}{\alpha}P$. That is, if a muscle pressure P is applied then the elastic artery contracts by an amount P/α. However, when blood is flowing the pressure of the blood varies dynamically; it is not constant.

Substituting into the governing equations and linearising as before, we obtain the same equations

$$\frac{\partial \hat{R}}{\partial t} + \frac{1}{2}R_*\frac{\partial \hat{v}}{\partial x} = 0 \quad \text{and} \quad \frac{\partial \hat{v}}{\partial t} = -\frac{\alpha}{\rho}\frac{\partial \hat{R}}{\partial x} - \frac{1}{\rho}\frac{\partial P}{\partial x} \,,$$

except that a new term from the muscular action has appeared. Eliminating $\hat{R}$ from these equations, as before, gives

$$\underbrace{\frac{\partial^2 \hat{v}}{\partial t^2} = c_*^2\frac{\partial^2 \hat{v}}{\partial x^2}}_{\text{as before}} - \underbrace{\frac{1}{\rho}\frac{\partial^2 P}{\partial x \partial t}}_{\text{forcing}} \,.$$

Observe that if P is constant in time, that is $P = P(x)$, or P is constant in space, that is $P = P(t)$, then $\frac{\partial^2 P}{\partial x \partial t} = 0$ and muscle activity has no effect upon the blood flow. It is only the combination of varying muscle activity in both x and t which does any useful work.

Experimentally, at any fixed location x the applied muscle pressure P depends upon t as shown in Figure 5.3(a). This is called a **twitch**. To have any effect, P must also depend upon x. One way to do this is to twitch at different locations at different times, as shown in Figure 5.3(b). Typically a *wave* of twitching occurs, passing with velocity w down the artery. That is, if the muscle at x_1 commences its twitch at time t_1

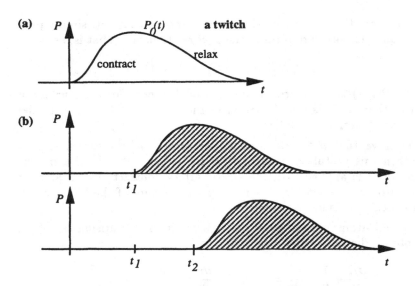

Figure 5.3: (a) a typical shape for a muscular twitch; (b) a twitch travels down the artery with some velocity w.

then a time Δt later the muscle at $x_1 + w\Delta t$ commences its twitch, and a time $2\Delta t$ later the muscle at $x_1 + 2w\Delta t$ commences its twitch, *etc.* Suppose that

$$P(x,t) = P_0(t - x/w) ,$$

where $P_0(t)$ is the muscle-applied pressure at $x = 0$, as shown in Figure 5.3(a). This describes a squeezing motion which travels down the artery with velocity w, and furthermore is the same at each location x.

Substituting this progressive squeezing into the modified wave equation gives

$$
\begin{aligned}
\frac{\partial^2 \hat{v}}{\partial t^2} - c_*^2 \frac{\partial^2 \hat{v}}{\partial x^2} &= -\frac{1}{\rho} \frac{\partial^2}{\partial x \partial t} P_0(t - x/w) \\
&= -\frac{1}{\rho} \frac{\partial}{\partial t} \left[P_0'(t - x/w) \frac{-1}{w} \right] \\
&= \frac{1}{\rho w} P_0''(t - x/w) .
\end{aligned}
$$

This guess works because every term in the DE is twice differentiated.

We already know a general homogeneous solution to this forced wave equation, it is $\hat{v} = f(x - c_* t) + g(x + c_* t)$. Now find a particular solution for this forcing: guess $\hat{v}(x,t) = K P_0(t - x/w)$ for some constant K.

Using this guess

$$\frac{\partial^2 \hat{v}}{\partial t^2} = K P_0''(t - x/w) \quad \text{and} \quad \frac{\partial^2 \hat{v}}{\partial x^2} = K \left(\frac{-1}{w}\right)^2 P_0''(t - x/w) ,$$

and the equation becomes

$$K P_0'' - \frac{K c_*^2}{w^2} P_0'' = \frac{1}{\rho w} P_0''$$

$$\Rightarrow \quad K \left(1 - \frac{c_*^2}{w^2}\right) = \frac{1}{\rho w}$$

$$\Rightarrow \quad K = \frac{1}{\rho w (1 - c_*^2/w^2)} = \frac{w}{\rho(w^2 - c_*^2)} .$$

Thus the general solution is

$$\hat{v} = \underbrace{\frac{w}{\rho(w^2 - c_*^2)} P_0(t - x/w)}_{\text{muscle forced flow}} + \underbrace{f(x - c_* t) + g(x + c_* t)}_{\text{free, elastic pulse}} , \tag{5.6}$$

which describes the forced flow due to the action of the muscles around the artery and also the free, elastic wave propagation along the artery.

Observe the factor $(w^2 - c_*^2)$ in the denominator of the forced flow. If the velocity at which the twitch travels, w, is approximately the same as the velocity of the elastic pulse, c_*, then the response to a given muscle pressure twitch is very large. That is, for best effect the muscle twitch should be made to travel with a velocity w which is close to c_*. Moreover, we want w to be just greater than c_* so that the factor $\frac{w}{\rho(w^2 - c_*^2)} > 0$ and a positive squeeze results in a positive contribution to the circulation of the blood.

5.2.3 The heart

We model the heart as a special case of an active artery. Consider an artery with a closed end at $x = 0$; then, as shown in Figure 5.4, the region near the closed end models the left ventricle of the heart. In fact in this model the heart is only distinguishable from the aorta by the presence of a valve, whose location we leave unspecified for the moment as we will fix it to be open for the time being.

Further suppose that there is no free, elastic pulse propagating to the left from the right, that is take $g = 0$ in the general solution which

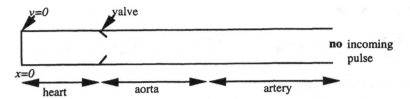

Figure 5.4: a portion of the heart–aorta–artery system drawn as a one-dimensional tube with active arteries.

was derived in the previous section. This is reasonable—it is hard to imagine how the small arteries or capillaries could generate a coherent pulse propagating up the artery towards the heart. Thus the general solution reduces to

$$\hat{v} = \frac{w}{\rho(w^2 - c_*^2)} P_0(t - x/w) + f(x - c_* t) \; .$$

But the closed end at $x = 0$ asserts that no blood can cross $x = 0$ and so there we must have $v = 0$. This allows us to determine the other arbitrary function f as follows

$$0 = \frac{w}{\rho(w^2 - c_*^2)} P_0(t) + f(-c_* t) \quad \text{upon putting } x = 0$$

$$\Rightarrow \quad f(-c_* t) = -\frac{w}{\rho(w^2 - c_*^2)} P_0(t)$$

$$\Rightarrow \quad f(X) = -\frac{w}{\rho(w^2 - c_*^2)} P_0(-X/c_*) \quad \text{upon setting } X = -c_* t$$

$$\Rightarrow \quad f(x - c_* t) = -\frac{w}{\rho(w^2 - c_*^2)} P_0(t - x/c_*) \; .$$

Thus the flow in the heart-artery system is

$$\hat{v} = \frac{w}{\rho(w^2 - c_*^2)} \Big\{ \underbrace{P_0(t - x/w)}_{\text{forced +ve flow}} - \underbrace{P_0(t - x/c_*)}_{\text{elastic -ve flow}} \Big\} \; .$$

Suppose $w > c_*$ (necessary for the heart to pump blood) then at any fixed location x the muscular contraction commences before the free, elastic pulse wave arrives, as shown in Figure 5.5. In this solution, the muscle forced outflow necessarily equals the backflow of the elastic response; there is no net movement of the blood! However, in the body this does not happen as the valves at the exit of the heart close and stop the backflow.

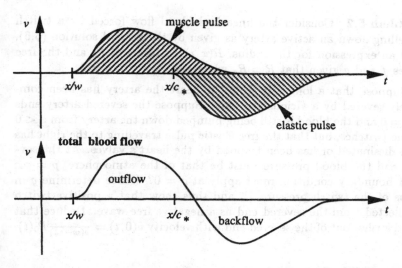

Figure 5.5: the blood velocity at any fixed location x as a superposition of the forced flow and the free, elastic wave.

We deduce from the above analysis is that the heart needs to be big enough so that, at the location of the valve, the muscle-forced pulse and the free-elastic pulse have separated in time. This is so the valves can catch all the outflow before the backflow starts. Quantitatively, if a twitch takes T seconds then, for the two pulses to just be separate at location x, we need $x/c_* - x/w = T$ and so the valve should be positioned at the location $x = Twc_*/(w - c_*)$.

Exercises

Problem 5.1 Consider the nonlinear dynamics of an inactive artery as MEDIUM modelled by equations (5.4–5.5). Show that both these partial differential equations become the same nonlinear unidirectional wave equation if we prescribe that the blood velocity is directly dependent upon R by the relation $v = V(R) = 2\sqrt{2\alpha/\rho}\left(R^{1/2} - R_*^{1/2}\right)$. Hence find the wave speed of "small" disturbances to the state of no blood flow, $v \approx 0$, in a resting artery, $R \approx R_*$. Suppose there was a pulse of blood coupled to an extension of the artery travelling down the artery according to this unidirectional wave equation ; would you expect the nonlinear dynamics to steepen the front of the pulse, or the back of the pulse?

MEDIUM   **Problem 5.2** Consider the linearised blood flow forced by a twitch
travelling down an active artery as given by the general solution (5.6).
Find an expression for the radius, $R(x,t)$, in terms of P_0 and the free
pulses f and g given that $R = R_*$ whenever $P_0 = f = g = 0$.

Suppose that a long way from the heart the artery has been completely severed by a vicious axe blow. Suppose the severed artery ends
at $x = 0$ and the blood is still being pumped down the artery from $x < 0$
by the twitches, but that the free elastic pulse travelling to the right has
been dissipated or has been trapped by the heart's valves. At the severed end the blood pressure must be that of the atmosphere, $p = p_*$;
what boundary condition must apply at $x = 0$? Hence determine g in
terms of the twitch pressure P_0 and thus show that a positive twitch
is reflected from the severed end as a negative free wave. Deduce that
blood gushes out of the severed end with velocity $\hat{v}(0,t) = \frac{1}{\rho(w-c_*)}P_0(t)$.

5.3   The flow of water

Consider the movement of water in a river or channel with a flat horizontal bottom and which is of constant breadth b. Let the height of the
water be $h(x,t)$ so that the varying height of the water surface makes
this a quasi-one-dimensional continuum, as shown in Figure 5.6. In common situations water is essentially of constant density. Also, the cross-
sectional area of the water in the channel is given by the breadth times
the height, $A(x,t) = b\,h(x,t)$, and thus the continuity equation (5.1)
gives

$$\frac{\partial}{\partial t}(bh\rho) + \frac{\partial}{\partial x}(bh\rho v) = b\rho\left[\frac{\partial h}{\partial t} + \frac{\partial(hv)}{\partial x}\right] = 0 \,,$$

which simplifies to

$$\frac{\partial h}{\partial t} + \frac{\partial(hv)}{\partial x} = 0 \,. \tag{5.7}$$

It should be mentioned that in a river, for example, the actual water
velocity is highly turbulent and varies significantly across a cross-section.
Thus, the one-dimensional velocity $v(x,t)$ which we use here is often
the cross-sectional *average* of the downstream velocity, rather than the
velocity at *every* point across the river.

The momentum equation (5.3) just needs to have the pressure field
specified. In order that Assumption 2 be satisfied, that *we may largely
ignore the curvature in the continuum*, it is necessary that the vertical
velocity of the water is quite small; as otherwise the surface $y = h(x,t)$

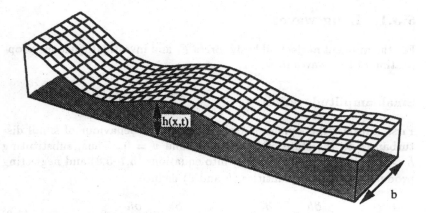

Figure 5.6: water of varying height $h(x,t)$ in a channel of breadth b

would have to curve significantly as the water moves up or down. Thus the pressure field within the water is approximately **hydrostatic**, that is, it increases linearly with depth below the water surface as described by

$$p = p_* + \rho g(h - y) ,$$

where p_* is the constant atmospheric pressure at the water's surface, and where y is height above the flat bottom. Substituting this into the momentum equation we immediately derive

$$\frac{\partial v}{\partial t} + v \frac{\partial v}{\partial x} + g \frac{\partial h}{\partial x} = \frac{F_1}{A\rho} . \qquad (5.8)$$

ASIDE

The two equations (5.7) and (5.8) contain the basic physical processes used in mathematical modelling of tides, for example. The equations describe the propagation of long waves (or tides), as investigated in Section 5.3.1. But to be more accurate they need to include body forces representing gravitational forcing and friction with the bottom, as described in Section 5.3.2 for rivers. Atmospheric forcing of the tides can easily be included by allowing the atmospheric pressure p_* to have a given dependence upon x and t. Of course tidal motion is typically two-dimensional, but the basic physics has been captured in this one-dimensional model.

5.3.1   Long waves

For the moment neglect all body forces F_1 and investigate just the propagation of free waves in water.

Small amplitude waves

The usual first step.   First *linearise* the equations by looking at the behaviour of small disturbances to the state of rest $h = h_*$ and $v = 0$. Thus, substituting $h = h_* + \hat{h}(x,t)$ and $v = \hat{v}(x,t)$ into equations (5.7–5.8) and neglecting products of "small' quantities ($\hat{h}$ and $\hat{v}$) deduce

$$\frac{\partial \hat{h}}{\partial t} + h_* \frac{\partial \hat{v}}{\partial x} = 0 \quad \text{and} \quad \frac{\partial \hat{v}}{\partial t} + g \frac{\partial \hat{h}}{\partial x} = 0 \;. \tag{5.9}$$

Differentiating the first of these with respect to t and eliminating $\frac{\partial \hat{v}}{\partial t}$ using the second gives

$$\frac{\partial^2 \hat{h}}{\partial t^2} = -h_* \frac{\partial^2 \hat{v}}{\partial x \partial t} = g h_* \frac{\partial^2 \hat{h}}{\partial x^2} \;.$$

This is the wave equation describing the propagation of disturbances, unchanged in form, with wave speed $c_* = \sqrt{g h_*}$. For example, tidal waves (tsunamis) in an ocean of depth about 1 km travel with a wave speed of $c_* = 100$ m/s (taking $g = 10$ m/s) which is $c_* = 360$ km/hr!

Unidirectional (simple) waves

Just as for ideal gas flow we readily find some exact solutions to the equations by looking for *unidirectional wave* solutions. Once again postulate that there exists a class of flows for which there is a simple functional relation between water height and water velocity, namely try the substitution $v = V(h)$ where $V(h)$ is some as-yet-unknown function. The continuity equation (5.7) becomes

$$\frac{\partial h}{\partial t} + V \frac{\partial h}{\partial x} + hV' \frac{\partial h}{\partial x} = 0 \;,$$

while the momentum equation (5.8) gives

$$V' \frac{\partial h}{\partial t} + VV' \frac{\partial h}{\partial x} + g \frac{\partial h}{\partial x} = 0$$

$$\Rightarrow \quad \frac{\partial h}{\partial t} + V \frac{\partial h}{\partial x} + \frac{g}{V'} \frac{\partial h}{\partial x} = 0 \;.$$

These two equations are equivalent only if

$$hV' = g/V'$$

$$\Rightarrow \left(\frac{dV}{dh}\right)^2 = \frac{g}{h}$$

$$\Rightarrow \frac{dV}{dh} = \pm\sqrt{\frac{g}{h}}$$

$$\Rightarrow V = \pm 2\sqrt{gh} + C \, ,$$

where C is a constant of integration. C is determined to be $\mp 2\sqrt{gh_*}$ by requiring that the water be in a state of rest, $v = 0$, when it is of depth $h = h_*$. Thus we have found two suitable relations for $v = V(h)$ which are substituted into either the continuity or momentum equations to give the class of exact solutions described by

$$v = \pm\left(2\sqrt{gh} - 2\sqrt{gh_*}\right) \quad \text{where} \quad \frac{\partial h}{\partial t} \pm \left(3\sqrt{gh} - 2\sqrt{gh_*}\right)\frac{\partial h}{\partial x} = 0 \, .$$
$$(5.10)$$

Appreciate that the $+$ alternative describes waves travelling to the right, and the $-$ alternative describes waves travelling to the left, by considering "small" waves on water of depth h_*. In this situation the evolution equation becomes simply $\frac{\partial h}{\partial t} \pm \sqrt{gh_*}\frac{\partial h}{\partial x} \approx 0$ which is the appropriate *linear* unidirectional wave equation.

Just as for gas dynamics the differential equation in (5.10) is of the form $\frac{\partial h}{\partial t} + c(h)\frac{\partial h}{\partial x} = 0$ which may be solved by the method of characteristics, as was done for car traffic in Section 2.2.4. As shown in Figure 5.7, the difference between car traffic and water waves is that here $c(h) = 3\sqrt{gh} - 2\sqrt{gh_*}$ (for waves travelling to the right) is monotonic increasing while in car traffic $c(\rho)$ was monotonic decreasing, compare with Figure 2.5. Thus the general conclusion that almost all variations in water height will eventually cause characteristics to cross and form "shocks" is still true. However, for water waves the shocks will form in a region of increasing height h (as opposed to occurring in a region of decreasing car density). Thus waves will break and form a turbulent bore on their front, as indeed is seen on every beach.

The Korteweg-deVries equation

> I was observing the motion of a boat which was rapidly
> drawn along a narrow channel by a pair of horses, when
> the boat suddenly stopped—not so the mass of water in the

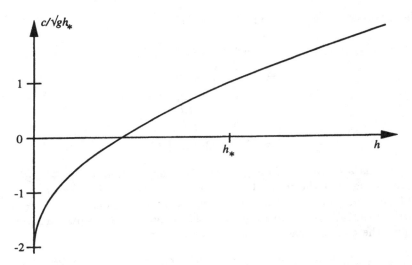

Figure 5.7: the *wave speed* of water of depth h when the water is stationary at a depth of h_*.

channel which it had put in motion; it accumulated round the prow of the vessel in a state of violent agitation, then suddenly leaving it behind, rolled forward with great velocity, assuming the form of a large solitary elevation, a rounded, smooth and well-defined heap of water, which continued its course along the channel apparently without change of form or diminuation of speed. I followed it on horseback, and overtook it still rolling on at a rate of some eight or nine miles an hour, preserving its original figure some thirty feet long and a foot to a foot and a half in hight. Its hight gradually diminished, and after a chase of one or two miles I lost it in the windings of the channel. Such, in the month of August 1834, was my first chance interview with that singular and beautiful phenomenon ... *John Scott Russell, 1844* [10].

At 13 or 14 kilometres per hour, 9 metres in length, and 30 to 50 centimetres in height.

The analysis of the previous subsection suggested that all waves on water will steepen and eventually break. Although wave breaking is endemic at beaches it is by no means the case that *all* water waves break, as observed so strikingly by Scott Russell. Water waves have a property called dispersion. Throw a stone into a pond or lake and observe the spreading group of waves.

If attention be fixed on a particular wave, it is seen to advance through the group, gradually dying out as it approaches the front, whilst its former place in the group is occupied in succession by other waves which have come forward from the rear. ... *John Scott Russell, 1844* [10, p369].

If wave-crests can appear and disappear then their energy must be travelling at a different velocity to that of the individual wave-crests, this is dispersion. In long water waves, the effect of dispersion is small, but it is enough to stop a wave breaking if the wave is not too large. The Korteweg-deVries equation describes unidirectional long water waves with dispersion taken into account.

Actually we do not yet have the requisite mathematical tools to properly derive this equation. However, some quick heuristic arguments end up with the correct equation. Consider first the linearised unidirectional wave equation on water of depth h_*, namely $\frac{\partial \hat{h}}{\partial t} + c_* \frac{\partial \hat{h}}{\partial x} = 0$ where $c_* = \sqrt{gh_*}$. The most basic wave motion is sinusoidal and so if we try $\hat{h} = A\cos(kx - \omega t)$, where k is known as the wavenumber of the wave and ω is known as the frequency, we find that $\omega = c_* k$, which is known as a **dispersion relation**. But, from some more advanced fluid mechanics it is known that the exact dispersion relation for small amplitude sinusoidal water waves in water of depth h_* is

$$\omega = \sqrt{gk\tanh(kh_*)} \approx c_* k - \frac{1}{6}c_* h_*^2 k^3$$

by a Taylor's expansion in either small h_* (shallow water) or small k (long waves). Thus the wave equation gives only a first approximation $\omega \approx c_* k$ to the correct dispersion relation. A second approximation is found by noting that a factor of k in the dispersion relation corresponds to a spatial derivative in the wave equation. Thus the modified wave equation

$$\frac{\partial \hat{h}}{\partial t} + c_* \frac{\partial \hat{h}}{\partial x} + \frac{1}{6}c_* h_*^2 \frac{\partial^3 \hat{h}}{\partial x^3} = 0 \,,$$

possesses a more appropriate dispersion relation for long water waves. This equation is similar to the nonlinear unidirectional wave equation (5.10), the differences being that here the nonlinearity has been ignored and that the dispersion term $\frac{1}{6}c_* h_*^2 \frac{\partial^3 \hat{h}}{\partial x^3}$ has been included. To include *both* the nonlinear and dispersion processes in the one equation we need only include this dispersion term into the nonlinear wave equation and arrive at

$$\frac{\partial \hat{h}}{\partial t} + \left(3\sqrt{g(h_* + \hat{h})} - 2\sqrt{gh_*}\right)\frac{\partial \hat{h}}{\partial x} + \frac{1}{6}c_* h_*^2 \frac{\partial^3 \hat{h}}{\partial x^3} = 0 \ .$$

Because of all the approximations done to date, we may as well make another one and expand the nonlinear term multiplying $\frac{\partial \hat{h}}{\partial x}$ in a Taylor's series in small $\hat{h}$. Thus we derive that long water waves travelling to the right in water of depth h_* evolve approximately according to

First derived in 1895.

$$\frac{\partial \hat{h}}{\partial t} + c_* \left(1 + \frac{3\hat{h}}{2h_*}\right)\frac{\partial \hat{h}}{\partial x} + \frac{1}{6}c_* h_*^2 \frac{\partial^3 \hat{h}}{\partial x^3} = 0 \ , \qquad (5.11)$$

known as the **Korteweg-deVries equation**. This equation involves nonlinearity which, as we have seen before, leads to wave breaking, and dispersion which may prevent this breaking.

ASIDE

> The preceding derivation is by no means rigorous, but it is flexible. It should be clear that the argument may be applied to any weakly dispersive wave system, and not just to water waves. Thus any dispersion relation $\omega(k)$ which is an odd function of k may be expanded in two terms to give a similar linear differential equation. Nonlinear terms are typically of the same form as appears in equation (5.11). In this way the Korteweg-deVries equation has arisen in many fields—plasma physics for example.

The solitary wave

Now, John Scott Russell observed a wave of constant shape travelling with constant velocity. This suggests that we look for a solution of the Korteweg-deVries equation in the form

$$\hat{h} = h_* \eta(X) \quad \text{where} \quad X = x - Ut$$

in which U is the velocity of translation and $\eta(X)$ is the shape of the wave. Substituting into equation (5.11), and using dashes to denote derivatives with respect to X, gives

$$-U\eta' + c_* \left(1 + \frac{3}{2}\eta\right)\eta' + \frac{1}{6}c_* h_*^2 \eta''' = 0$$

$$\Rightarrow \quad \frac{1}{6}h_*^2 \eta''' + \frac{3}{2}\eta\eta' - \left(\frac{U}{c_*} - 1\right)\eta' = 0$$

$$\Rightarrow \quad \frac{1}{6}h_*^2 \eta'' + \frac{3}{4}\eta^2 - \left(\frac{U}{c_*} - 1\right)\eta + D = 0 \quad \text{upon integrating}$$

$$\Rightarrow \quad \frac{2}{3}h_*^2\eta''\eta' + 3\eta^2\eta' - 4\left(\frac{U}{c_*} - 1\right)\eta\eta' + 4D\eta' = 0 \quad \text{multiply by } 4\eta'$$

$$\Rightarrow \quad \frac{1}{3}h_*^2{\eta'}^2 + \eta^3 - 2\left(\frac{U}{c_*} - 1\right)\eta^2 + 4D\eta = E \quad \text{upon integrating}$$

where D and E are arbitrary constants of integration. One of the appealing features of this last equation is that it is analogous to the equation of a particle moving in a potential well, and the solutions may thus easily be understood. The analogy goes as follows: η represents the location of the "particle"; X is a time-like variable and so η' is the velocity of the "particle"; $\frac{1}{3}h_*^2{\eta'}^2$ then represents the kinetic energy while $V(\eta) = \eta^3 - 2\left(\frac{U}{c_*} - 1\right)\eta^2 + 4D\eta$ represents the shape of potential energy in the well; and finally E is the total energy of the "particle".

The useful aspect of the analogy is that by simply drawing some potential wells, $V(\eta)$, for various values of the parameters U and D (see Figure 5.8), we easily visualise the possible solutions. For example, sinusoidal oscillations take place in a parabolic potential well—thus **sinusoidal** wave solutions correspond to *small* oscillations in the bottom of the local minimum of the potential well. Larger oscillations in the potential well, obtained by increasing the "energy" E say, will still correspond to water waves; it is just that they are no longer purely sinusoidal, they are deformed by the nonlinearity, $\hat{h}\frac{\partial \hat{h}}{\partial x}$, in the Korteweg-deVries equation.

One intriguing wave solution is found by imagining a particle which is nearly, but not quite, balanced on the local maximum of $V(\eta)$, as seen in Figure 5.8. At first the particle will hardly move, as it will seem to be balanced; but it will slowly start to slip off the maximum, gaining speed as it moves away, until it rolls rapidly down the potential well, up the other side, reaches its maximum distance to the right and then falls back down the potential well, up the left hand side, slowing down as it rolls up to the maximum, and eventually stops precariously balanced on top of the local maximum. Now, remember that in this analogy "time" is actually X and the particle's "position" η is actually the height of the water surface. Thus, in this particular solution, the starting and finishing "position" of the particle is actually the height of the water surface a long way upstream and downstream; and this is the height which we want to be the undisturbed water height $h = h_*$ which is $\eta = 0$. Hence, choose $D = 0$ so that the location of the maximum of the potential is at $\eta = 0$, as seen in Figure 5.8, and choose the "energy" $E = 0$ to obtain the above described solution which is illustrated in Figure 5.9. This wave consists of a single hump of raised water, whose

The solitary wave.

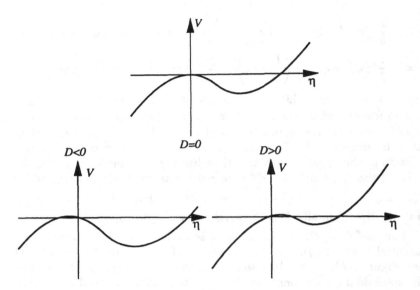

Figure 5.8: Some potential wells $V(\eta)$ for solutions to the Korteweg-deVries equation for $U > c_*$ and various values of D.

shape progresses along with constant velocity, it is indeed the **solitary wave** as described by Scott Russell.

An analytic expression for the shape of the solitary wave is found either by guesswork or by elementary integration. With $D = E = 0$ the differential equation for $\eta(X)$ becomes

$$\frac{1}{3}h_*^2 \left(\frac{d\eta}{dX}\right)^2 + \eta^3 - \alpha\eta^2 = 0 \ ,$$

where $\alpha = 2(U/c_* - 1)$ is a measure of the amplitude of the wave, see Figure 5.9. This equation is separable and becomes

$$\int \frac{d\eta}{\eta\sqrt{\alpha - \eta}} = \frac{\sqrt{3}}{h_*} \int dX \ .$$

One of the standard substitutions to eliminate the square root in the integrand is $\eta = \alpha\mathrm{sech}^2(\theta)$ which eventually gives the solution to be

$$\eta = \alpha\mathrm{sech}^2 \left(\sqrt{\frac{3\alpha}{4h_*^2}}X\right) \ .$$

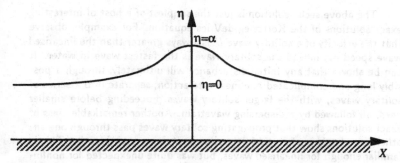

Figure 5.9: a solitary wave on water: it translates without change of shape.

Denoting the wave amplitude by $A = \alpha h_*$ and writing this solution in terms of the original variables, the solitary wave is described by

$$\hat{h} = A\operatorname{sech}^2\left[\sqrt{\frac{3A}{4h_*^3}}(x - Ut)\right] , \qquad (5.12)$$

which travels at the velocity $U = c_*\left(1 + \frac{A}{2h_*}\right)$. This is an exact solution of the Korteweg-deVries equation for all amplitudes A. However, the Korteweg-deVries equation is only approximate, it is derived under the assumption that A/h_* is much less than 1, and in fact solitary waves in water are found to peak at a maximum height of $A/h_* \approx 0.7$.

> This is a most beautiful and extraordinary phenomenon: the first day I saw it was the happiest day of my life. Nobody had ever had the good fortune to see it before or, at all events, to know what it meant. It is now known as the solitary wave of translation. No one before had fancied a solitary wave as a possible thing. When I described this to Sir John Herschel, he said "It is merely half of a common wave that has been cut off". But it is not so, because the common waves go partly above and partly below the surface level; and not only that but its shape is different. Instead of being half a wave it is clearly a whole wave, with this difference, that the whole wave is not above and below the surface alternately but always above it. So much for what a heap of water does: it will not stay where it is but travels to a distance. ... *John Scott Russell, 1865* [11, p208].

The above sech² solution is just the simplest of a host of interesting exact solutions of the Korteweg-deVries equation. For example, observe that the velocity of a solitary wave U is always greater than the linearised wave speed c_*; indeed the solitary wave is the fastest wave in water. It can be shown that any initial disturbance will ultimately, through a possibly long and complicated nonlinear interaction, separate into a series of solitary waves, with the larger solitary waves proceeding before smaller ones, all followed by a dispersing wavetrain. Another remarkable class of exact solutions show that propagating solitary waves pass through one another completely unchanged in shape and retaining their identity. This is familiar enough for linearised waves, but was quite unexpected for nonlinear waves, such as those described by the Korteweg-deVries equation: the complicating nonlinearity would generally destroy the identity of the waves through the timespan of their interaction. However, it is now known that quite a few nonlinear wave systems exhibit this "nice" interaction between solitary waves.

5.3.2   River flow

A river generally slopes down gently to the sea and the water flows because gravity pulls the water down, the larger the slope of the river the faster the flow. To quantify this effect, note that if a river slopes down with angle α then gravity has a component along the continuum, which provides a body force of $F_1 = A\rho g \sin\alpha$ on the water. The slope is usually very small and so we approximate $\sin\alpha$ by α.

Frictional resistance stops the water from accelerating indefinitely. This resistance primarily comes from friction with every point across the bottom of the river and is thus proportional to the breadth b. It is found that the resistance is also proportional to the square of the local velocity. Thus it provides a drag component in the body force of $F_1 = -C\rho b v^2$ where C is a drag coefficient which depends on geometric features of the river, such as the roughness and the detailed shape of the bottom.

Including both of these effects in the momentum equation (5.8) gives

$$\frac{\partial v}{\partial t} + v\frac{\partial v}{\partial x} + g\frac{\partial h}{\partial x} = \frac{F_1}{A\rho} = g\alpha - C\frac{v^2}{h} \ . \tag{5.13}$$

This is to be solved together with the continuity equation (5.7).

Note that the kinematic approximation as used in Problem 3.6 is obtained by simply neglecting the left-hand-side of the above momentum equation, which then gives $v = \sqrt{g\alpha/C}h^{1/2}$ in order that the gravitational pull balances the frictional drag. This leads immediately to the Chezy law as used in the problem. If, in addition, the depth of water h is constant along the river, then this balance is the **uniform stream** solution of the two equations; it describes a river flowing smoothly down to the sea, without any waves or other disturbances.

Stability: roll waves

Such a uniform stream of water may or may not be stable. To investigate the possible behaviour, linearise the equations and look at the behaviour of "small" perturbations to the uniform state $h = h_*$ and $v = v_*$, where

$$Cv_*^2 = g\alpha h_* \ .$$

As before, substitute $h = h_* + \hat{h}(x, t)$ and $v = v_* + \hat{v}(x, t)$ and neglect products of the small hatted quantities. The continuity and momentum equations then become

$$\frac{\partial \hat{h}}{\partial t} + v_* \frac{\partial \hat{h}}{\partial x} + h_* \frac{\partial \hat{v}}{\partial x} \ = \ 0$$

$$\frac{\partial \hat{v}}{\partial t} + v_* \frac{\partial \hat{v}}{\partial x} + g \frac{\partial \hat{h}}{\partial x} + g\alpha \left(\frac{2\hat{v}}{v_*} - \frac{\hat{h}}{h_*} \right) \ = \ 0 \ .$$

Multiply the second equation by h_* and differentiate with respect to x, then use the first equation to eliminate $\hat{v}$ and so give

$$-\frac{\partial}{\partial t} \left(\frac{\partial \hat{h}}{\partial t} + v_* \frac{\partial \hat{h}}{\partial x} \right) - v_* \frac{\partial}{\partial x} \left(\frac{\partial \hat{h}}{\partial t} + v_* \frac{\partial \hat{h}}{\partial x} \right) + gh_* \frac{\partial^2 \hat{h}}{\partial x^2} +$$

$$+ g\alpha \left[-\frac{2}{v_*} \left(\frac{\partial \hat{h}}{\partial t} + v_* \frac{\partial \hat{h}}{\partial x} \right) - \frac{\partial \hat{h}}{\partial x} \right] = 0$$

$$\Rightarrow \frac{\partial^2 \hat{h}}{\partial t^2} + 2v_* \frac{\partial^2 \hat{h}}{\partial x \partial t} + (v_*^2 - gh_*) \frac{\partial^2 \hat{h}}{\partial x^2} + \frac{2g\alpha}{v_*} \left[\frac{\partial \hat{h}}{\partial t} + \frac{3v_*}{2} \frac{\partial \hat{h}}{\partial x} \right] = 0 \ .$$

Just as for amoebae in Section 2.3, we wish to investigate the behaviour of disturbances of arbitrary wavelength on this stream. Because of the presence of derivatives of different order in the above equation the analysis here is a little more complicated; physically these complications arise from the fact that disturbances travel as waves on the stream as well as growing or decaying. The easiest way to proceed is to look for solutions of the form

$$\hat{h} = \Re\{A\exp(ikx + st)\} ,$$

where A is a measure of the initial amplitude of the disturbance of wavenumber k, that is of wavelength $2\pi/k$, and where the real part of s, $\Re\{s\}$, is the growth-rate of the wave and the imaginary part of s, $\Im\{s\}$, is the frequency of the wave. Substitute this into the above linearised equation and deduce that

$$\Re\left\{A\exp(ikx + st)\left[s^2 + 2iv_*ks - \left(v_*^2 - gh_*\right)k^2 + \right.\right.$$
$$\left.\left. + \frac{2g\alpha}{v_*}\left(s + \frac{3v_*}{2}ik\right)\right]\right\} = 0$$
$$\Rightarrow \quad s^2 + 2iv_*ks - \left(v_*^2 - gh_*\right)k^2 + \frac{2g\alpha}{v_*}\left(s + \frac{3v_*}{2}ik\right) = 0$$

since the equation has to hold for all x and t. Now, this is a not-so-simple quadratic equation for $s(k)$, and if the real part of s is ever positive for some wavenumber k then the uniform stream will be unstable.

However, the above equation was derived under the assumption that the waves or disturbances of the water surface were relatively long; thus it is only valid for "small" values of the wavenumber k. This observation simplifies the analysis at this point—we simply solve this equation iteratively in a scheme which is appropriate for small k. Write it as

$$s = -\frac{3v_*}{2}ik + \frac{v_*}{2g\alpha}\left[\left(v_*^2 - gh_*\right)k^2 - 2iv_*ks - s^2\right] ,$$

and start the iteration by substituting the guess $s \approx 0$ into the right-hand side. The first nontrivial approximation is then

$$s \approx -\frac{3v_*}{2}ik ,$$

but this has zero real part and so predicts that the waves just travel, they do not grow or decay. However, substituting this first approximation into the right-hand side gives the second approximation to be

Each iteration gives one more term in the series in k.

$$s \approx -\frac{3v_*}{2}ik + \frac{v_*}{2g\alpha}\left[\frac{1}{4}v_*^2 - gh_*\right]k^2 .$$

Observe that the real part of s, although small for small k (of size k^2), is definitely non-zero and so waves on a flowing stream generally either grow or decay. Whether they grow or decay depends upon the sign of the square bracketed expression in the above equation. If $v_* > 2\sqrt{gh_*}$ then the real part of s is positive and long waves are unstable. That is, if the free stream velocity v_* is greater than twice the wave velocity $\sqrt{gh_*}$ on stationary water then long waves will grow in amplitude, destroying the uniform flow. Equivalently, this instability will occur only if $\alpha > 4C$ (from the relation between h_* and v_*); that is, only if the slope is larger than four times the non-dimensional drag coefficient of the channel.

In rivers v_* is typically much less than $\sqrt{gh_*}$, but man-made channels and spillways easily exceed the critical values, usually because they are smooth and have a low drag coefficient. The flow resulting from such an instability often takes the form of **roll waves** as shown in Figure 5.10. This sort of flow has a roughly periodic sequence of discontinuous turbulent bores (or "shocks") separated by smoothly varying profiles. During rain storms it may be sometimes seen in the water which is flowing along the road surface.

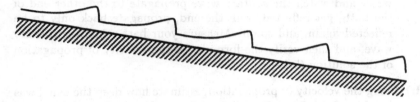

Figure 5.10: roll waves on a sloping surface.

More information about river flow, flood waves and glacier flow may be found in the book on waves by Whitham [12, §3.2–3]; in which there is also much more on long water waves and solitary waves in §13.10–15.   ASIDE

Exercises

Problem 5.3 Consider a rectangular bay of constant width b, of con-   EASY
stant depth h_*, and extending from the ocean's edge at the mouth of the bay at $x = 0$ to vertical cliffs at $x = L$. Inside the bay the dynamics of small amplitude tidal motions are described by equation (5.9) The tidal motion in the deep ocean moves so much water that in effect it specifies the water elevation at the mouth of the bay; hence take the boundary condition there to be that $\hat{h} = a\sin\omega t$ at $x = 0$. a is the amplitude of the tides and ω is their frequency. No water can flow through the cliffs at $x = L$; what is the boundary condition there? Using separation of

variables, find the small amplitude particular solution for $\hat{h}$ and $\hat{v}$ (that is, the solution proportional to a). For some tidal frequencies, ω, the response in the bay is apparently very large; why?

MEDIUM   **Problem 5.4** Using the same approach as in Problem 5.3 investigate the small-amplitude water flow in a canal which connects two oceans at $x = 0$ and $x = L$. Suppose the tidal motions in each ocean are of the same frequency but have different amplitudes and phases.

EASY   **Problem 5.5** It is quite easy to generate solitary waves.

(a) Generate a solitary wave in a bath—as the sloping bottom complicates the phenomenon, a container with a flat bottom is preferable. Fill a bath with water to a depth of 5–10 cms and let the water motion settle down. At one end, push a small heap of water forward with your hands and forearms, take your hands out of the water and watch the solitary wave propagate to the other end of the bath, get reflected from the end, propagate back only to be reflected again, and so on. Measure your bath, time the solitary wave, and then verify the formula for the velocity of propagation of the solution (5.12).

(b) Using the velocity of propagation, estimate how deep the canal was in which John Scott Russell first observed the solitary wave.

(c) Solitary waves can also propagate along "vortex tubes". Get a cup of coffee and stir it so that the coffee circulates around the cup with a patch of froth spinning on top in the middle of the cup. In the centre of the cup, stretching from the bottom to the top of the coffee, is a vortex tube. Briefly insert a teaspoon into the top of the coffee in the middle of the patch of froth; this will nearly stop the froth from revolving but upon taking out the teaspoon the froth will speed up to spin with the rest of the coffee. Now watch carefully and observe that every second or so the patch of froth will briefly slow down in its spin and then speed up again! The explanation of this curious phenomenon is that the brief insertion of the teaspoon into the top of the vortex tube generated a localised slowing down of the vortex tube, and this localised slowing down travelled down the vortex tube, just like a hump of water in the bath, until it reached the bottom, is reflected off the bottom and travels back up to the surface where we see it slowing down the froth once again. This solitary wave

A bucket sized cup makes things easier to see.

will continue to travel up and down the vortex tube in the cup for a little time.

Problem 5.6 Consider the minus case of equation (5.10) which describes exact solutions corresponding to left-travelling waves. Use the method of characteristics to solve the **dam break** problem in which a dam located at $x = 0$ holding back stationary water of depth (height) $h = h_*$ suddenly and catastrophically breaks at time $t = 0$. *Hint:* this is just like a queue of cars at a traffic light. At what speed does the waterfront travel downstream? What is the height of water at the location of the ex-dam as time passes?

MEDIUM

As seen in Alastair McLean's *Force 10 from Navarone*

Problem 5.7 Modify the momentum equation (with $F_1 = 0$) by the inclusion of a term in $\frac{\partial^3 h}{\partial x^3}$ so that, together with the unmodified continuity equation (5.7), *small* amplitude waves $A\cos(kx - \omega t)$ on water of depth h_* more closely match the exact dispersion relation for water waves of $\omega^2 = gk\tanh(kh_*)$. You will have derived a form of the **Boussinesq equations** describing the motion of shallow water.

MEDIUM

Unlike the Korteweg-deVries equation which only describes wave motion travelling in one direction, the Boussinesq equations are a complete description, although approximate, of the motion of long waves in water.

ASIDE

Chapter 6

Applications of 1-D continuum mechanics

One-dimensional continuum mechanics helps our understanding of a wide variety of physical situations. To appreciate just some of the possibilities we here describe briefly some such applications. These descriptions are intentially brief and informal; they are designed to give just a flavour of the applications.

6.1   Dispersion in a pipe, river or channel

Imagine some material being carried along by the flow of water in a river, channel or pipe. Whether the material is natural or a contaminant, dissolved or in suspension we shall term it a **tracer**, and assume that it has a negligible effect on the dynamics of the fluid which is transporting it. Some examples would be: salt or nitrates in a river or estuary; a spill of pollutants into a drain; chemicals in a flow reactor.

The simplest model of the evolution of the tracer (whatever the tracer may be) is that it is carried downstream at the average velocity

141

of the flow. This can be seen by considering any one tracer particle. The natural turbulence in the stream, or perhaps random molecular motion, will carry the particle randomly to and fro across the stream as it is swept downstream. Thus as time proceeds, on average the tracer particle will "experience" the average downstream velocity of the flow of water. Let $C(x, t)$ denote the one-dimensional density of the tracer material, that is its concentration in units of *amount of material per distance downstream*, and let $U(x, t)$ denote the average velocity of the stream at any cross-section x and time t.

The tracer is conserved and so the continuity equation (2.2) must apply, with $r = 0$ if there are no generation or decay mechanisms. By the above reasoning the tracer is carried with a flux $q = UC$ and so

$$\frac{\partial C}{\partial t} + \frac{\partial (UC)}{\partial x} = 0$$

is the model equation for the evolution of the tracer. If, in addition, the average velocity is constant then the above differential equation becomes simply the one-dimensional wave equation

$$\frac{\partial C}{\partial t} + U \frac{\partial C}{\partial x} = 0$$

with general solution

$$C(x, t) = C_0(x - Ut)$$

where $C_0(x)$ is the initial concentration of the tracer. Thus, as for linear waves, this model would predict that the tracer travels downstream at the mean velocity of the stream, maintaining the shape of its distribution.

However, the above model is not satisfactory. In practice we find that a localised pulse of tracer will spread significantly as it travels downstream—there does exist an along-stream dispersive mechanism. This along-stream spreading may be very much larger than could be accounted for by molecular diffusion or turbulent mixing. How can we model it?

The simplest way to treat spreading along the stream, in the x direction, is to treat it as a diffusive process (as explained in Section 2.3.1); that is, we assume the flux of tracer is not UC but is instead

$$q = UC - D \frac{\partial C}{\partial x} ,$$

which turns the continuity equation (2.2) into the form of an **advection-diffusion equation**:

$$\frac{\partial C}{\partial t} + \frac{\partial (UC)}{\partial x} = \frac{\partial}{\partial x}\left(D\frac{\partial C}{\partial x}\right) . \tag{6.1}$$

However, the unresolved question is: what is the correct value for the effective diffusion constant D? Until it can be calculated the above model is useless.

The mechanism of this downstream spreading is that, although on average each tracer particle experiences the average downstream velocity as it randomly wanders across the stream, it is also the case that through natural random fluctuations some tracer particles will spend longer than average in the slow-flowing sides of the stream, and thus be retarded, while other tracer particles will spend longer than average in the faster-flowing core of the stream, and thus be carried further downstream. The resultant spreading of the tracer particles is effectively a downstream diffusion. To estimate the coefficient D of this diffusion we need to know more details about the distribution of tracer across the stream and how this affects the evolution.

ASIDE

In this aspect it is rather like the analysis of the bending of a beam (Section 4.2.2), where the internal distribution of stress causes an along-beam distribution of bending moment which is vital to the beam's ability to support a transverse load.

To show how to find D in a simple example, consider the two-dimensional stream shown in Figure 6.1. The concentration of tracer is $c(x, y, t)$ in units of quantity per unit area. The stream flows along the channel with a downstream velocity $u(y)$ which only varies across the stream. The cross-stream mixing or molecular diffusion of tracer is represented with a diffusion constant κ. These two physical processes are all that are necessary to show, as in Figure 6.1, an enhanced diffusion downstream. Take it as given that the relevant equation governing the detailed evolution of c is in this case

$$\frac{\partial c}{\partial t} + u(y)\frac{\partial c}{\partial x} = \kappa\frac{\partial^2 c}{\partial y^2} \tag{6.2}$$

where the tracer cannot diffuse out of the stream and so

$$\frac{\partial c}{\partial y} = 0 \quad \text{at} \quad y = \pm b$$

where $y = \pm b$ are the sides of the channel.

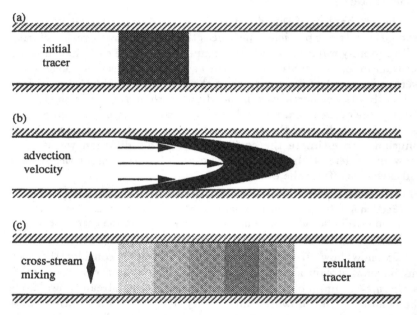

Figure 6.1: One cycle of physical effects to show how along-stream dispersion may arise: (a) an initial slug of tracer; (b) slug is advected downstream differentially; (c) and then is mixed across-stream to effectively spread the tracer along the stream.

The above equation is quite difficult to solve directly, as it involves two spatial dimensions. Our aim is to find the coefficients of the considerably simpler one-dimensional model equation (6.1). Now an effective downstream diffusion has an effective flux of $-D\frac{\partial C}{\partial x}$ and so the dispersion is linked to downstream gradients of concentration. An exact solution, involving downstream gradients, of the governing equation (6.2) is found by trying $c = \alpha[x - Ut + v(y)]$; this is a distribution of tracer with a uniform gradient α, moving downstream with a velocity U, and which has some cross-stream structure $v(y)$. Substituting into (6.2) and dividing by α requires

$$\kappa v'' = -U + u(y) \qquad \text{such that} \qquad v' = 0 \quad \text{at} \quad y = \pm b .$$

For any given velocity profile $u(y)$ this is a simple second-order differential equation for $v(y)$. For example, if the velocity profile is parabolic $u(y) = \frac{3U}{2}(1 - y^2/b^2)$ then $v(y) = \frac{-Ub^2}{120\kappa}\left(9 - 30(y/b)^2 + 5(y/b)^4\right)$.

Given such an exact solution, that $c = \alpha[x - Ut + v(y)]$, the crucial part of the analysis is the following heuristic argument. Propose that relatively soon after the release of any tracer, the concentration of tracer is sufficiently smooth so that in any smallish reach of the stream the above exact solution will approximately hold for some value of the gradient α.

Upon neglecting some technical difficulties

- The component of the concentration in the local view, $\alpha[x - Ut]$, corresponds, in the large, to the distribution of the tracer along the stream, namely the cross-stream average $C(x, t)$.

- Further, the gradient α of the tracer in the local view corresponds, in the large view, to the gradient $\frac{\partial C}{\partial x}$.

Thus the above exact solution transforms to the approximate balance that $c \approx C + v(y)\frac{\partial C}{\partial x}$. This equation simply describes the approximate details of the tracer distribution c whenever there is a large-scale variation in the cross-sectional average concentration C.

To derive a differential equation for C, simply substitute this into the governing equation (6.2), and take the cross-sectional average of the resulting equation to obtain

$$\frac{\partial C}{\partial t} + \frac{\partial}{\partial x}\left[UC - D\frac{\partial C}{\partial x}\right] = 0 \qquad \text{where} \quad D = -\overline{u(y)v(y)} \qquad (6.3)$$

where the overbar denotes the cross-sectional average. This is simply the one-dimensional advection-diffusion equation (6.1) which we argued earlier was needed. However, we now have an expression for the effective diffusion coefficient D in terms of the velocity profile in the stream. For example, if the velocity profile is parabolic, as specified earlier, then $D = \frac{2U^2b^2}{105\kappa}$.

The utility of this achievement is that in order to predict how a tracer is to evolve along a stream we need only solve this relatively simple one-dimensional, albeit approximate, evolution equation rather than the much more complicated two-dimensional (or three-dimensional in actual practice) evolution equation (6.2). This one-dimensional approximation is used in an enormous variety of physical situations.

ASIDE
> The derivation of the above approximation has been via some very shaky heuristic arguments. Nonetheless, they are the basis of some vitally important models of processes in the environment. With the aid of some more mathematical techniques this approximation may be made rigorously [8]. Such a revision is important as only then can we show the strengths and the limitations of such practical approximations.

6.2   Solidification of a binary alloy

When a pure liquid is supercooled to a temperature below its freezing point, then the solidification front between the frozen solid and the supercooled liquid becomes extremely convoluted. Snowflakes are a common example of the possible intricate branching patterns which may ensue. When the liquid is a mixture of two or more components, called an **alloy**, then the formation of such regions of **dendrites** during solidification is commonplace. We here discuss a simple model [13] for the region of dendritic growth. The model treats the region of mixed solid and liquid as a continuum whose properties depend upon the relative proportions of liquid and solid.

For most applications this model is appropriate because it is the overall properties of the mixed solid–liquid matrix which is more important than the precise details of the microscopic structure of the dendrites. The region of mixed phases, solid dendrites and liquid mix, is termed the **mush** or the **mushy zone**. Treating the mush as a new phase of the continuum matter, distinct from either solid or liquid, we follow the evolution of two phase boundaries as the alloy solidifies: the solid/mush interface; and the mush/liquid interface. However, these new phase boundaries are geometrically simple planes, rather than convoluted, and the resultant governing one-dimensional equations become tractable. The equations are based on conservation principles applied on a macroscopic scale which is much larger than the microscopic scale of the dendrites.

The phase diagram

The dynamics of solidification hinge upon the equilibrium phase diagram for the particular system under consideration. For simplicity, we use here a common idealisation.

Consider a system comprising two chemically distinct components A and B, for example, water and salt. The mix is either a liquid or a solid depending upon the temperature T and the relative concentrations of the components. Let C be the concentration of the component B (salt), and so the concentration of the component A (water) is inversely dependent upon C. The concentration and temperature of the mix correspond to a point in the (C, T)-plane, and whether it is a liquid or a solid is indicated in Figure 6.2: for temperatures above the **liquidus**, $T = -\mu C$, the mix is a liquid; for temperatures below the **solidus**, $T = T_E$ (the eutectic temperature), the mix is a solid; the mix cannot exist at concentrations and temperatures in the shaded region between the liquidus and the solidus; and this figure does not apply to concentrations above the **eutectic concentration**, C_E. If a liquid mix of A and B (salt water) is cooled then when it eventually evolves to the liquidus it cannot cool any further as a *mix* as it would then enter the forbidden region. We find that the component A (water) solidifies as a pure material (pure ice), as if the state of the mix had jumped to the left, vertical branch of the solidus, and the component B (salt) is pushed out of the solid (ice) to diffuse into the liquid.

This is why frozen sea water is quite fresh

The dendrites: a mushy zone

Consider the situation shown in Figure 6.3 where a liquid binary mix is cooled from the bottom. In the region near the bottom ($x = 0$) the mix has solidified (frozen), but it is virtually pure component A (ice) as the component B (salt) has been pushed out during the solidification. This expulsion of B (salt) raises its concentration (salinity) near the growing solid and this extra material must diffuse away vertically into the liquid binary mix (salt water). However, along the liquidus in Figure 6.2 an increase in concentration means a decrease in the temperature at which solidification occurs. So it may happen that the liquid mix just a little way from the solid front, where the concentration of B (salinity) is less, is actually supercooled, that is it is locally below the liquidus in Figure 6.2.

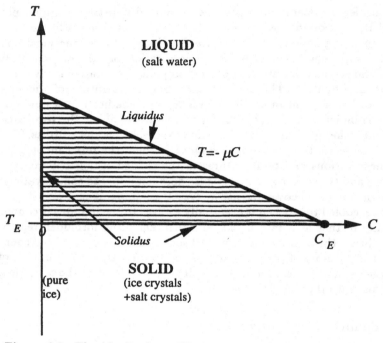

Figure 6.2: The idealised equilibrium phase diagram for a two component chemical system; the shaded region is forbidden.

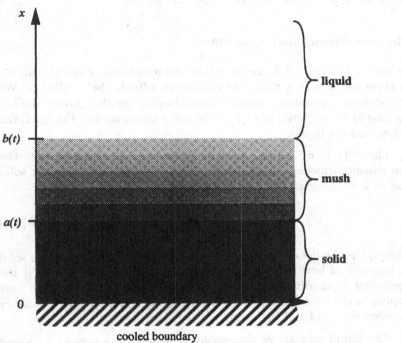

Figure 6.3: A schematic diagram of a liquid binary mix being cooled from the bottom so that it gradually solidifies.

In such a case the solidification actually occurs as thin dendrites (each of scale a millimetre thick), growing rapidly into the super-cooled region; these dendrites being separated by liquid. It is impossible to track the detailed shape of the increasingly convoluted solid–liquid boundary; instead we model the region of dendrites and liquid as a **mushy** zone, as shown in Figure 6.3, with properties in between that of the liquid and the solid. For example, on a scale much larger than the width of the dendrites, the mush can have average concentrations and temperature which lie in the forbidden region in Figure 6.2.

The one-dimensional equations

As shown in Figure 6.3, consider that the properties of the solidifying mixture depends only upon the vertical coordinate, here called x. We now deduce appropriate one-dimensional equations that govern the behaviour of the material in each of the three layers shown: the solid, the mush, and the liquid.

The only thing which varies in the solid is its temperature—the component B having been rejected. Heat can only diffuse in the solid and so must obey an equation of the form

$$\rho_s C_{p_s} \frac{\partial T}{\partial t} = k_s \frac{\partial^2 T}{\partial x^2} \,, \tag{6.4}$$

where ρ_s is the density of the solid, C_{p_s} is the specific heat of the solid (a measure of how well the solid can hold heat energy), and k_s is the coefficient of thermal conductivity (its diffusion). This equation only applies in the solidified region, $x < a(t)$, where $x = a(t)$ is the boundary between the solid and the dendritic, or mushy, region.

The liquid mix above the mushy zone has both temperature and concentration of solute (component B) varying. We assume that the liquid does not start moving as a result of density variations induced by the compositional and temperature variations, because otherwise the problem becomes formidable. In the absence of motion in the liquid, heat and solute can only diffuse through the liquid mix. Thus they also must each satisfy a simple diffusion equation, say

$$\rho_\ell C_{p\ell} \frac{\partial T}{\partial t} = k_\ell \frac{\partial^2 T}{\partial x^2} \tag{6.5}$$

$$\frac{\partial C}{\partial t} = D \frac{\partial^2 C}{\partial x^2} \tag{6.6}$$

where ρ_ℓ is the density of the liquid, C_{p_ℓ} is the specific heat of the liquid, k_ℓ is the thermal conductivity in the liquid, and D is the diffusivity of the solute B. These equations only apply in the liquid region ahead of the mushy zone, namely for $x > b(t)$, where $x = b(t)$ is the boundary between the liquid mix and the tip of the dendrites.

More complicated equations apply in the mushy zone, $a(t) < x < b(t)$. There the effective properties of the mush depend upon the ratio of liquid mix to solid dendrite in any small region of the mush. Measure this ratio by the fraction of volume taken by the liquid in such a small region, and denote it by χ; the volume fraction χ is a function of position in the mushy zone, and also of time. The fraction of volume taken up by the solid in any small region of the mush is then $1 - \chi(x, t)$. Once again assume that heat and solute are transported only by diffusion. However, there are other physical processes active in the mush; as the material solidifies latent heat is released and the solute is rejected. Thus the conservation equation for the temperature takes the form

$$(\rho C_p)_m \frac{\partial T}{\partial t} = \frac{\partial}{\partial x} \left(k_m \frac{\partial T}{\partial x} \right) - \rho_s L \frac{\partial \chi}{\partial t}, \qquad (6.7)$$

where

$$(\rho C_p)_m = \chi \, \rho_\ell C_{p_\ell} + (1 - \chi) \, \rho_s C_{p_s} \qquad (6.8)$$

is the effective density and specific heat of the mush in any locale,

$$k_m = \chi \, k_\ell + (1 - \chi) \, k_s \qquad (6.9)$$

is the effective thermal conductivity in the mush, and $-\rho_s L \frac{\partial \chi}{\partial t}$ represents the source of latent heat which is released as the liquid solidifies. The solute can only diffuse through the liquid and so, by conservation principles, its concentration C must satisfy an equation of the form

$$\chi \frac{\partial C}{\partial t} = \frac{\partial}{\partial x} \left(D\chi \frac{\partial C}{\partial x} \right) - C \frac{\partial \chi}{\partial t}, \qquad (6.10)$$

where $-C \frac{\partial \chi}{\partial t}$ represents the solute rejected by the solidifying material and expelled into the liquid mix.

Boundary conditions

To be complete we have to provide appropriate boundary conditions for the above equations. At the top of the liquid and at the bottom of the solid it is reasonable to specify that the temperature and the concentration of the solute is some known constant. However, we also have to supply boundary conditions at the two interfaces: the solid–mush interface $x = a(t)$; and the mush–liquid interface $x = b(t)$. The discussion of these conditions are nontrivial as they involve a moving boundary; we refer the interested reader to the paper by Worster [13].

Some typical solutions and profiles of the mush composition are also given there. We leave the trail here as the aim was simply to show how a one-dimensional model could be derived to analyse such a problem with such a complicated physical microstructure—the dendrites.

6.3   The greenhouse effect

Human beings' apparently unceasing discharge of waste gases into the atmosphere affects the ability of this layer of gas to protect us from the cold of interstellar space. Computer models to simulate the dynamics of the atmosphere, in an attempt to predict quantitatively the effect that our actions have on the atmosphere, are becoming increasingly sophisticated. Most of these atmospheric models have involved just one space dimension, the vertical height from the surface of the earth. It is only very recently that people have seriously been able to investigate two-dimensional models that resolve not only the vertical dynamics but also the variations with latitude—for centuries we have relied on one-dimensional models.

In this section we examine the balance between incoming radiation from the sun and the outgoing infrared radiation of the warm earth. The discourse is based on Ch. 2 of the book by Houghton [6] in which much of the basic dynamics of the atmosphere is discussed. The atmosphere plays a vital role in insulating the earth and any compositional changes in the air may disastrously affect this delicate balance. The indications are that our current habits will cause significant warming of the earth's surface because they enhance the greenhouse effect.

The first result to be discussed arises from the fact that the atmosphere is all but transparent to visible light. Because the sun radiates most of its energy in the visible spectrum, the atmosphere has virtually no effect on the sun's incoming radiation to the earth; the influence of the sun is, in effect, to heat up the surface of the earth with some known flux of energy. The only way that this energy can be lost from planet earth is via infrared radiation into space. This is where the atmosphere influences the heat budget. Air is opaque to infrared radiation—but this opacity is not due to the major constituents of nitrogen and oxygen, it is due to minor constituents such as carbon dioxide. The infrared radiation which is emitted by the warm ground is absorbed and re-emitted by the air many times before it can escape the atmosphere. Thus the atmosphere acts like a blanket to keep the earth's environment warm. The effectiveness of this blanket is determined by the dynamics of the infrared radiation.

Absorption and emission

There are three principle components in the interaction between the atmosphere and infrared radiation: the radiation travelling upwards; the radiation travelling downwards; and the vertical temperature structure of the atmosphere. First look at the vertically travelling radiation.

Let z measure distances from the ground, $z = 0$, to the top of the atmosphere, $z = L$. Infrared photons travelling vertically at the speed of light, $c \approx 3 \times 10^8$ m/s, are absorbed by the air at some average rate proportional to the density of the air and to the number of photons. Once absorbed, they heat the air and may be emitted in either direction by the black body radiation of the air. The infrared photons involved in these processes are generally of varying frequencies and so it is useful to frame the model in terms of the energy density of upward travelling photons, $E^{\uparrow}$, and the energy density of downward travelling photons, $E^{\downarrow}$. Actually, we will eventually deal only with the corresponding (unsigned) fluxes of infrared energy, namely $F^{\uparrow\downarrow} = cE^{\uparrow\downarrow}$.

The conservation of infrared photons travelling upwards leads to the principle of conservation of upward-travelling infrared energy. Thus the conservation equation (2.2) gives

$$\frac{\partial E^{\uparrow}}{\partial t} + \frac{\partial F^{\uparrow}}{\partial z} = -(\text{absorption}) + (\text{emission})$$
$$= -k\rho F^{\uparrow} + k\rho\pi B(T) \, ,$$

where $\rho(z)$ is the density of air, k is an absorption coefficient, and $B(T)$ is proportional to the upward-emitted black-body radiation due to the air being locally at a temperature T; $B \propto T^4$. However, $E^{\uparrow} = \frac{1}{c}F^{\uparrow}$ and thus may be neglected in this equation as c is very large. In essence, the speed of light is so very fast compared to the phenomena of interest (light takes about 10 μsec to traverse the atmosphere), that the vertical structure of the infrared energy density reaches equilibrium with the air temperature almost instantaneously. Thus the upward flux of infrared energy satisfies the equation

$$\frac{\partial F^\uparrow}{\partial z} = -k\rho \left(F^\uparrow - \pi B \right) \ . \tag{6.11}$$

Similar arguments lead to an analogous equation for the downward flux of infrared energy:

$$-\frac{\partial F^\downarrow}{\partial z} = -k\rho \left(F^\downarrow - \pi B \right) \ , \tag{6.12}$$

with the difference that the sign on the left-hand side is negative to correspond to the downwards direction of the radiative transport.

The third component of this model is the temperature structure of the atmosphere. The energy of absorbed infrared radiation is stored as heat in the air until it is emitted as black body radiation. In the absence of large-scale vertical motions of the atmosphere, which is certainly not true everywhere but suffices for an initial analysis, the heat energy of density $\rho c_p T$ is not transported and the flux is zero. Thus the principle of conservation of heat leads to the following equation

$$
\begin{aligned}
\rho c_p \frac{\partial T}{\partial t} &= \text{(total absorption)} - \text{(total emission)} \\
&= k\rho \left(F^\uparrow + F^\downarrow \right) - k\rho 2\pi B(T)
\end{aligned}
$$

which by using equations (6.11–6.12) becomes

$$\rho c_p \frac{\partial T}{\partial t} = -\frac{\partial}{\partial z} \left(F^\uparrow - F^\downarrow \right) \ . \tag{6.13}$$

The three equations (6.11–6.13) form the basic model for the interaction between infrared radiation and the temperature structure of the atmosphere.

Radiative equilibrium

Now consider the idealisation of a steady flux of energy in visible light from the sun, ϕ, heating up the ground at $z = 0$ to a temperature T_g. The ground warms until the ground is warm enough that it loses enough heat in infrared radiation through the atmosphere to exactly balance the incoming energy flux.

In this state of equilibrium the temperature T does not vary in time and so the left-hand side of equation (6.13) is zero. Integrating the right-hand side then gives that the net upwards transport of energy through radiation is

$$F^\uparrow - F^\downarrow = \text{a constant}.$$

In equilibrium this upwards loss of energy must exactly match the influx of visible energy from the sun and so

$$F^\uparrow - F^\downarrow = \phi \ .$$

Furthermore, let the total amount of radiative flux at any height be $\psi(z) = F^\uparrow + F^\downarrow$; then adding equations (6.11–6.12) gives

$$\frac{d\phi}{dz} = -k\rho \left(\psi - 2\pi B\right) \ .$$

But ϕ is a constant, so the left-hand side is zero and thus

$$\psi = 2\pi B \ .$$

In essence this equation asserts that the total flux of radiation at any height z is in balance with the black body radiation at that height. In addition, subtracting (6.12) from (6.11) leads to

$$\frac{d\psi}{dz} = -k\rho\phi \ .$$

Integrating this and eliminating ψ in favour of B gives

$$B(T) = -\frac{\phi}{2\pi} \int k\rho \, dz + \text{constant}$$

It is this last equation which describes, implicitly through $B(T)$, the temperature structure of the atmosphere. However, there is a constant in the equation which must be determined by a boundary condition. At the top of the atmosphere there is no gas above to radiate infrared light downwards; thus the boundary condition is $F^\downarrow = 0$ at $z = L$. Hence, at the top of the atmosphere $\psi = F^\uparrow = \phi$, which requires that $B = \phi/(2\pi)$ at $z = L$, which leads to

$$B(T) = \frac{\phi}{2\pi} \left[1 + \int_z^L k\rho \, dz' \right] \tag{6.14}$$

as governing the temperature structure.

At the bottom of the atmosphere, near $z = 0$, the air temperature, T_0, is such that the corresponding black body radiation B_0 is

$$B_0 = \frac{\phi}{2\pi} \left[1 + \chi_0 \right] \ ,$$

where $\chi_0 = \int_0^L k\rho \, dz$ is called the **optical depth** of the atmosphere.
However, the ground itself is at a somewhat higher temperature, T_g,
corresponding to black body radiation B_g. One way to see that this
must be so is to realise that the very top of the earth has to supply *all*
the infrared radiation travelling upwards from $z = 0$; at a level in the
air above, the radiation from the air below the level supplies some of the
upwards radiation to supplement the black body radiation of the air at
the level. The ground has to be hotter in order for it to supply all the
required radiation. The condition to determine B_g is that the upwards
flux of radiation energy at $z = 0$ is $F^\uparrow = \pi B_g$. However, in the air, and
in particular the air just above the ground,

$$
\begin{aligned}
2F^\uparrow &= \phi + \psi \qquad \text{by definition of } \phi \text{ and } \psi \\
&= \phi + 2\pi B_0 \qquad \text{since } \psi = 2\pi B.
\end{aligned}
$$

Using the earlier expression for B_0 we then obtain that the ground
temperature is such that

$$
B_g = \frac{\phi}{2\pi} \left[2 + \chi_0 \right] . \tag{6.15}
$$

The greenhouse effect

The optical depth χ_0 is a direct measure of the insulating effect of the
atmosphere. If $\chi_0 = 0$, that is there is no atmosphere, then $B_g =
\phi/\pi$ and the ground temperature is in radiative equilibrium with the
incoming radiation from the sun. If χ_0 is large then the ground and
lower atmosphere temperature will be considerably warmer.

The optical depth of the atmosphere depends primarily upon minor
constituents such as water vapour or carbon dioxide. If these increase
then so will the optical depth and the insulating effect of the atmosphere
will grow, causing the ground and the lower atmosphere to consequently
warm up.

Potentially, warming due to an increase in water vapour may be a
runaway effect. As the ground warms, more water evaporates which
increases the load of water vapour in the air, which then warms the
ground, and so on. This runaway process may explain why the planet
Venus became so hot.

Bibliography

[1] C.M. Bender & S.A. Orszag, *Advanced Mathematical Methods for Scientists and Engineers*, McGraw-Hill (1978).

[2] R. Haberman, *Mathematical Models: Mechanical Vibrations, Population Dynamics, and Traffic Flow*, Prentice-Hall (1977).

[3] P.G. Hodge Jr., *Continuum Mechanics: an Introductory Text For Engineers*, McGraw-Hill (1970).

[4] Y.C. Fung, *Foundations of Solid Mechanics*, Prentice-Hall (1965).

[5] Y.C. Fung, *A First Course in Continuum Mechanics*, 2nd edition, Prentice-Hall (1977).

[6] J.T. Houghton, *The Physics of Atmospheres*, 2nd edition, Cambridge University Press (1986).

[7] C.C. Lin & L.A. Segel, *Mathematics Applied to Deterministic Problems in the Natural Sciences*, Macmillan (1974).

[8] G.N. Mercer & A.J. Roberts, "A centre manifold description of contaminant dispersion in channels with varying flow properties," *SIAM J. Appl. Math.* **50** (1990) pp. 1547–1565.

[9] *New Scientist*, 16 April 1987.

[10] J. Scott Russell, "Report on waves," *Brit. Ass. Rep.* (1844).

[11] J. Scott Russell, *The Modern System of Naval Architecture. Vol. 1*, Day & Son, London (1865).

[12] G.B. Whitham, *Linear and Nonlinear Waves*, John Wiley & Sons (1974).

[13] M.G. Worster, "Solidification of an alloy from a cooled boundary,"
     J. Fluid Mechanics **167** (1985), pp. 481–501.

Index